Android® Phones

for dummies®
A Wiley Brand

4th edition

by Dan Gookin

for
dummies®
A Wiley Brand

Android® Phones For Dummies®, 4th Edition

Published by: **John Wiley & Sons, Inc.,** 111 River Street, Hoboken, NJ 07030-5774, www.wiley.com

Copyright © 2016 by John Wiley & Sons, Inc., Hoboken, New Jersey

Published simultaneously in Canada

For general information on our other products and services, please contact our Customer Care Department within the U.S. at 877-762-2974, outside the U.S. at 317-572-3993, or fax 317-572-4002. For technical support, please visit https://hub.wiley.com/community/support/dummies.

Wiley publishes in a variety of print and electronic formats and by print-on-demand. Some material included with standard print versions of this book may not be included in e-books or in print-on-demand. If this book refers to media such as a CD or DVD that is not included in the version you purchased, you may download this material at http://booksupport.wiley.com. For more information about Wiley products, visit www.wiley.com.

Library of Congress Control Number: 2016952319

ISBN: 978-1-119-31068-6; 978-1-119-31071-6 (ebk); 978-1-119-31072-3 (ebk)

Manufactured in the United States of America

10 9 8 7 6 5 4 3 2 1

Table of Contents

Introduction

t may be a smartphone, but it makes you feel dumb. Don't worry: You aren't alone. As technology leaps ahead, it often leaves mortal humans behind. You paid good money for your phone — why not use all of its features?

This book makes the complex subject of Android phones understandable. It's done with avuncular care and gentle handholding. The information is friendly and informative, without being intimidating. And yes, ample humor is sprinkled throughout the text to keep the mood light.

About This Book

I implore you: Do not read this book from cover to cover. This book is a reference. It's designed to be used as you need it. Look up a topic in the table of contents or the index. Find something about your phone that vexes you or something you're curious about. After getting the answer, get on with your life.

Every chapter in this book is written as its own self-contained unit, covering a specific Android phone topic. The chapters are further divided into sections representing tasks you perform with the phone or explaining how to get something done. Sample sections in this book include

>> Typing without lifting a finger

>> Making a conference call

>> Dealing with a missed call

>> Uploading a picture to Facebook

>> Recording video

>> Creating a mobile hotspot

>> Flying with your phone

>> Extending battery life

Every section explains a topic as though it's the first one you read in this book. Nothing is assumed, and everything is cross-referenced. Technical terms and topics, when they come up, are safely shoved to the side, where they're easily avoided. The idea here isn't to learn anything. This book's philosophy is to help you look it up, figure it out, and move on.

How to Use This Book

This book follows a few conventions for using your phone, so pay attention!

The main way to interact with an Android phone is by using its touchscreen, which is the glassy part of the phone as it's facing you. The physical buttons on the phone are called keys. These items are discussed and explained in Part I of this book.

Various ways are available to touch the screen, which are described in Chapter 3.

Chapter 4 covers typing text on an Android phone, which involves using something called the onscreen keyboard. When you tire of typing, you can dictate your text. It's all explained in Chapter 4.

This book directs you to do things on your phone by following numbered steps. Every step involves a specific activity, such as touching something on the screen; for example:

1. **Choose Downloads.**

This step directs you to tap the text or item on the screen labeled Downloads. You might also be told to do this:

1. **Tap Downloads.**

Because this book covers a variety of phones, alternative commands may be listed. One of them is bound to match something on your phone, or at least be close to what you see:

1. **Tap the My Downloads action or the Downloads action.**

Various phone settings can be enabled or disabled, as indicated by a master control, which looks like the On/Off toggle shown in the margin. Slide the button to the right to activate the switch, turning on a phone feature. Slide the button to the left to turn off the feature.

Foolish Assumptions

Even though this book is written with the gentle handholding required by anyone who is just starting out or is easily intimidated, I have made a few assumptions.

I'm assuming that you're still reading the introduction. That's great. It's much better than getting a snack right now or checking to ensure that the cat isn't chewing through the TV cable again.

My biggest assumption: You have an Android phone. It can be any Android phone from any manufacturer supported by any popular cellular service provider in the United States. Because Android is an operating system, the methods of doing things on one Android phone are similar, if not identical, to doing things on another Android phone. Therefore, one book can pretty much cover the gamut of Android phones.

Android has versions. This book was updated to cover the current Android release, 6.0, called Marshmallow. Also addressed is Android 5.1, known as Lollipop. Some details on older versions of Android phones might be found here as well. These are all similar versions of the operating system, so if your phone has an older version, you should be just fine.

To confirm which Android version your phone has, follow these steps.

1. **At the Home screen, tap the Apps icon.**

2. **Open the Settings app.**

3. **Choose the About Phone item.**

   On some Samsung phones, you need to first tap the General tab atop the screen and then swipe down the screen to find an About Device item. Samsung phones can be a little different from other Android phones, and those differences are highlighted throughout this tome.

4. **Look at the item titled Android Version.**

   The version number is listed, such as 6.0.1.

Don't fret if these steps confuse you: Review Part I of this book, and then come back here. (I'll wait.)

More assumptions: You don't need to own a computer to use your Android phone. If you have a computer, great. The Android phone works well with both PCs and Macs. When phone and computer cross paths, you'll find directions for both PC and Mac.

Finally, this book assumes that you have a Google account. If you don't, find out how to configure one in Chapter 2. Having a Google account opens up a slew of useful features, information, and programs that make using your phone more productive.

Icons Used in This Book

TIP

This icon flags useful, helpful tips or shortcuts.

REMEMBER

This icon marks a friendly reminder to do something.

WARNING

This icon marks a friendly reminder not to do something.

TECHNICAL STUFF

This icon alerts you to overly nerdy information and technical discussions of the topic at hand. Reading the information is optional, though it may win you a pie slice in Trivial Pursuit.

Beyond the Book

The publisher maintains a support page with updates or changes that occur between editions of this book. Go to www.dummies.com, search for *Android Phones For Dummies,* then open the Extras tab on this book's specific page to view the updates or changes. Or click the Cheat Sheet link to view helpful information pulled from throughout the text.

On a personal note, my email address is dgookin@wambooli.com. Yes, that's my real address. I reply to every email I receive, especially when you keep your question short and specific to this book. Although I enjoy saying "Hi," I cannot answer technical support questions, resolve billing issues, or help you troubleshoot your phone. Thanks for understanding.

My website is wambooli.com. This book has its own page on that site, which you can check for updates, new information, and all sorts of fun stuff. Visit often:

```
wambooli.com/help/android/phones
```

Where to Go from Here

Thank you for reading the introduction. Few people do, and it would save a lot of time and bother if they did. Consider yourself fortunate, though you probably knew that.

Your task now: Start reading the rest of the book — but not the whole thing, and especially not in order. Observe the table of contents and find something that interests you. Or look up your puzzle in the index. When these suggestions don't cut it, just start reading Chapter 1.

1

Getting Started with Your Android Phone

IN THIS PART . . .

Get started with Android phones.

Work through the setup of your Android phone.

Learn how to use your Android phone.

Discover parts of the Android phone.

Chapter 1

Hello, Phone!

t may have a funky name, like a character in a science fiction novel or a sports hero. Or it can simply be a fancy number, perhaps with the letter X thrown in to make it cool. No matter what, the phone you own is an *Android* phone because it runs the Android operating system. The adventure you're about to undertake begins with removing the thing from the box and getting to know your new smartphone.

Liberation and Setup

The phone works fastest when you remove it from its box. How you liberate it is up to you. I prefer to gingerly open the box, delicately lifting the various flaps and tenderly setting everything aside. I even savor the industrial-solvent smell. If you prefer, you can just dump everything on the tabletop. Be careful: Your phone may be compact, but it's not indestructible.

ANDROID PHONE-BUYING TIPS

When buying a phone, first look at a cellular provider, and then determine which phones are available and suit your purpose.

Finding a cellular provider is all about coverage: Can you get a signal everywhere you need one? Despite the boasts, not every cellular provider offers full data coverage. The true test is to ask people who frequent your same locations which services they use and whether they're happy with the coverage.

All Android phones offer similar features and a vast array of apps. Start looking for a phone by finding ones that feel good in your hands. Some people like smaller, compact phones that fit easily in a pocket or purse. Others prefer the large-format (phablet) phones, which offer larger screens.

Check the phone's display, not by reviewing the fancy technical jargon but by looking at it with your own eyes. View some photos on the phone to see how good they look.

Phones come with varying quantities of storage, from 8GB on up to 128GB and more. Some phones might still offer removable storage in the form of a microSD, though this feature is becoming rather rare.

Camera resolution isn't vital, but if your Android phone is your only digital camera, getting a high-resolution rear camera is a plus.

Beyond these basic items, most Android phones are drearily similar. To work best with this book, ensure that your phone uses the Android operating system and can access and use Google Play, the online Android store, where you obtain apps, music, video, and books. Some low-price, bargain phones restrict your purchases to the manufacturer's own app store. That's not a good thing.

Several useful items are found loitering inside your Android phone's box. Some of them are immediately handy, and others you should consider saving for later. Even if you've already opened the box and spread its contents across the table like some sort of tiny yard sale, take a few moments to locate and identify these specific items:

>> The phone itself, which may be fully assembled or in pieces

>> Papers, instructions, a warranty, and perhaps a tiny, useless Getting Started pamphlet

>> The phone's back cover, which might already be attached to the phone

>> The charger/data USB cable

» The charger head, which is a wall adapter for the charger/data cable

» Other stuff, including the SIM card, SIM card removal tool, earbuds, carrying case, or other goodies

It's rare, but the phone may feature a removable battery. If so, you'll find the phone's battery in the box, along with the phone's rear cover. These items must be assembled. Look for instructions inside the box.

If anything is missing or appears to be damaged, immediately contact the folks who sold you the phone.

TIP

I recommend keeping the instructions and other information as long as you own the phone: The phone's box makes an excellent storage place for that stuff — as well as for anything else you don't plan to use right away.

See the later section "Adding accessories" for a description of various goodies available for the typical Android phone.

Phone Assembly

Most Android phones come fully assembled. If yours doesn't, the folks at the Phone Store have most likely put everything together for you. When you're on your own, some setup may be required. This process might involve installing the SIM card or microSD card and inserting the battery. Directions that come with the phone assist you.

Don't worry about the assembly process being overly complex; if you're good with Legos, you can put together an Android phone.

Removing the plastic sheeting

The phone ships with a clingy plastic sheeting over its screen, back, or sides. The sheeting might tell you where to find various features, so look it over before you peel it off. And, yes, you need to remove the sheeting; it's for shipping protection, not for long-term phone protection.

TIP

» Remove all clingy plastic sheets.

» Check the phone's rear camera to confirm that you've removed the plastic sheeting from its lens.

» Feel free to throw away the plastic sheeting.

Installing the SIM card

A *SIM card* identifies your phone on a digital cellular network. Before you can use the phone, the SIM card must be installed. The only time you as a mere mortal need to do this is when you purchase the phone independently of a cellular provider. Otherwise, the kind people at the Phone Store install the SIM card. They pretend like it's a task that requires a PhD, but installing a SIM card is simple.

When you do need to install the SIM card yourself, follow these steps when the phone is turned off:

1. Pop the SIM card out of the credit-card-size holder.

Push the card with your thumb and it pops out. Don't use scissors or you may damage the card.

2. Locate the SIM card cover on the phone's outer edge.

The cover features a dimple or hole on one end.

3. Insert the SIM card removal tool into the hole on the SIM card cover; press it in firmly.

The SIM card cover pops up or the SIM card tray slides out.

4. Insert the SIM card into the SIM card slot, or place the SIM card into the SIM card tray and reinsert it into the phone.

The SIM card is shaped in such a way that it's impossible to insert improperly. If the card doesn't slide into the slot, reorient the card and try again.

5. Close the SIM card cover.

You're done.

The good news is that you seldom, if ever, need to remove or replace a SIM card.

>> On some phones, the SIM card is inserted internally. In that case, remove the phone's back cover and, if necessary, remove the battery to access the SIM card slot.

>> SIM stands for subscriber identity module. SIM cards are required for GSM cellular networks as well as for 4G LTE networks.

TECHNICAL
STUFF

Installing a microSD card

A few Android phones offer removable storage in the form of a microSD card. If your phone sports this feature, go out and obtain a microSD card to take advantage of the extra storage.

No, your phone didn't come with a microSD card — unless the Phone Store included it as a "bonus." (You still paid for it.)

To insert a microSD card, heed these directions:

1. **Locate the slot into which you stick the microSD card.**

   The slot is labeled as shown in Figure 1-1. It is not the same as the SIM card slot.

microSD card hatch or cover

FIGURE 1-1:
Opening
the microSD
card hatch.

Lift here

2. **Flip open the teensy hatch on the microSD card slot.**

   Insert your thumbnail into the tiny slot on the hatch. Flip the hatch outward. It's attached on one end, so it may not completely pop off.

3. **Insert the microSD card into the slot.**

   The card goes in only one way. If you're fortunate, a little outline of the card illustrates the proper orientation. If you're even more fortunate, your eyes will be good enough to see the tiny outline.

TIP

You may hear a faint clicking sound when the card is fully inserted. If you don't, use the end of a paperclip or your fingernail to fully insert the card.

>> It's okay to insert the microSD card while the phone is on.

>> If the phone is on, a prompt appears on the touchscreen, detailing information about the card. The card is available instantly for use. If not, the card may need to be formatted. See Chapter 18 for details.

>> Some older Android phones may feature internal microSD card slots. In that case, you must remove the phone's back cover to access and install the card. You might even have to remove the battery to get at the card.

>> I've never seen an Android phone come with a microSD card. If your phone can use such a card, obtain one at any computer or office supply store. They're cheaper if you order them on the Internet.

>> A microSD card comes in a capacity rated in gigabytes (GB), just like most media storage or memory cards. Common microSD card capacities are 8GB, 16GB, 32GB, and higher. The maximum size allowed in your phone depends on its design. The side of the phone's box lists compatible capacities.

Removing the microSD card

To remove the microSD card, follow these steps:

1. Turn off the phone.

It's possible to remove the card while the phone is on, and directions are offered in Chapter 18. For now, ensure that the phone is off. Specific power-off directions are found in Chapter 2.

2. Open the little hatch covering the microSD card slot.

Refer to the preceding section.

3. Using your fingernail or a bent paperclip, gently press the microSD card inward a tad.

The microSD card is spring-loaded, so pressing it in pops it outward.

4. Pinch the microSD card between your fingers and remove it completely.

After you've removed the card, you can continue using the phone. It works just fine without a microSD card.

WARNING

>> A microSD card is teensy! If you remove it from your phone, keep it in a safe place where you won't lose it. Never stick the microSD card into your ear.

>> You can purchase microSD card adapters to allow a computer to read the card's data on a computer. The adapter allows the microSD card to insert into a standard SD memory slot or the USB port.

>> Refer to Chapter 18 for more information on phone storage.

Charge the Battery

The phone's battery may have enough oomph in it to run the setup-and-configuration process at the Phone Store. If so, count yourself lucky. Otherwise, you need to charge the phone's battery. Don't worry about flying a kite and waiting for a lightning storm. Instead, follow these steps:

1. **If necessary, assemble the charging cord.**

   Connect the charger head (the plug thing) to the USB cable that comes with the phone.

2. **Plug the charger head and cable into a wall socket.**

3. **Plug the phone into the USB cable.**

   The charger cord plugs into the micro-USB connector, found at the phone's bottom.

As the phone charges, you may see a charging–battery graphic on the touchscreen, or a notification lamp on the phone's front side may glow. Such activity is normal.

The phone may turn on when you plug it in for a charge. That's okay, but read Chapter 2 to find out what to do the first time the phone turns on.

>> I recommend fully charging the phone before you use it.

>> Older USB cables use the micro-A connector, which plugs in only one way. If the cable doesn't connect to the phone, flip over the cable and try again.

>> Newer USB Type-C cables and connectors plug in any which way.

>> You can use the phone while it's charging, although the phone won't turn on when the battery charge is too low.

>> The phone also charges itself whenever it's connected to a computer's USB port. The computer must be on for charging to work. Some phones may charge only when plugged into powered USB ports, such as those found directly on the computer console.

>> Cell phones charge more quickly when plugged into the wall than into a computer's USB port or a car adapter.

>> Unlike with the old NiCad batteries, you don't need to worry about fully discharging your phone before recharging it. If the phone needs a charge, even when the battery is just a little low on juice, feel free to do so.

>> Some Android phones can be charged wirelessly. See the later section "Adding accessories."

>> Also see Chapter 23 for battery and power management information.

Android Phone Orientation

No one told the first person to ride a horse which way to sit. Some things just come naturally. Your Android phone most likely isn't one of those things. It requires a special introduction and orientation.

Finding things on your phone

I think it's cute when people refer to things that they can't name as *doodads* or *thingamabobs*. Cute, but inaccurate. Take a gander at Figure 1-2, which illustrates common items found on the front and back of a typical Android phone.

Not every item shown in the figure may be in the exact same spot on your phone. For example, the Power/Lock key might be found on the top of the phone, not the side.

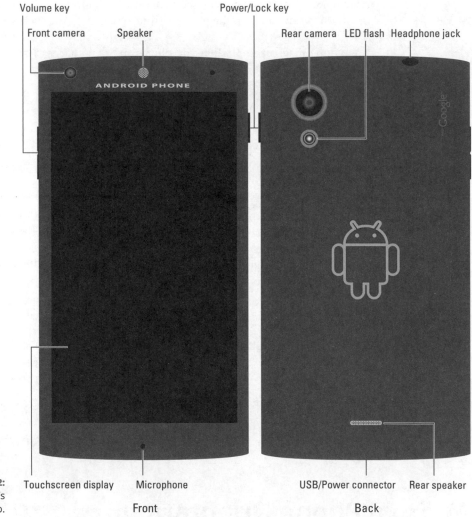

FIGURE 1-2: Your phone's face and rump.

Front

Back

The terms referenced in Figure 1-2 are the same as the terms used elsewhere in this book and in whatever scant Android phone documentation exists. Here are the highlights:

Power/Lock key: The Power/Lock key does more than turn the phone on or off, which is why it's the Power/Lock key and not the On/Off button.

Volume key: The phone's volume control is two buttons in one. Press one end of the key to set the volume higher; the other end sets the volume lower. This key might also be used to control the phone's camera, as covered in Chapter 13.

Touchscreen display: The main part of the phone is its *touchscreen* display. It's a see-touch thing: You look at the display and also touch it with your fingers to control the phone. That's where it gets the name *touch*screen.

Front camera: The phone's front-facing camera is found above the touchscreen. It's used for taking self-portraits as well as for video chat.

Speaker(s): The primary phone speaker is located top center on the phone. One or more additional speakers might also be found on the phone's bottom edge or backside.

Microphone: Somewhere below the touchscreen, you'll find the microphone. It's tiny, about the diameter of a pin. Don't stick anything into the hole! A second, noise-canceling microphone might also be found on the back of the phone.

Headphone jack: On the phone's top or bottom edge, you'll find a hole where you can connect standard headphones.

Rear camera: The rear camera is found on the phone's back. It may be accompanied by one or two LED flash gizmos.

USB/power connector: Use this important connector to attach the phone's USB cable. That cable is used to charge the phone and to communicate with a computer. This connector might be on the side of the phone, but more commonly it's found bottom center.

Take a moment to locate all items mentioned in this section, as well as shown in Figure 1-2, on your own phone. It's important that you know where they are.

>> Additional Items you might want to locate on your phone, items not illustrated in Figure 1-2, include the SIM card cover and microSD card slot. Use of these features is covered earlier in this chapter.

>> Some phones, such as those in the Samsung Galaxy line, feature a physical button called the Home key. This key is found below the touchscreen. On some Samsung phones, the physical Home key also serves as a fingerprint reader.

- » The Galaxy Note line of phones features a pointing device, in the form of a digital stylus called an S Pen. It docks at the phone's bottom edge.

- » It's common for some LG and other phones to feature controls on the back. You may find the power button near the top center on the back of the phone, a volume key, or a fingerprint reader.

- » The phone's microphone picks up your voice, loud and clear. You don't need to hold the phone at an angle for the microphone to work.

Using earphones

You can use your Android phone without earphones, such as the common earbuds, but they are certainly a nice thing to have. If the nice folks who sold you the phone tossed in a pair of earphones, that's wonderful! If they didn't, well then, they weren't so nice, were they?

The earbud–style earphone sets directly into your ears. The sharp, pointy end of the earphones, which you don't want to stick into your ear, plugs into the phone's headphone jack.

Between the earbuds and the sharp, pointy thing, you might find a doodle button. The button is used to answer a call, mute the phone, start or stop the playback of music, and perform other functions.

A teensy hole on the doodle serves as a microphone. The mic allows you to wear the earbuds and talk on the phone while keeping your hands free. If you gesture while you speak, you'll find this feature invaluable.

- » If your Android phone didn't come with a set of earbuds, you can purchase a pair at any electronics store where the employees wear name tags. Ensure that the earbuds feature a microphone or doodle button.

REMEMBER

- » Be sure to fully insert the earphone connector into the phone. The person you're talking with can't hear you well when the earphones are plugged in only part of the way.

- » You can also use a Bluetooth headset with your phone, to listen to a call or some music. See Chapter 17 for more information on Bluetooth.

TIP

- » Fold the earphones when you put them away, as opposed to wrapping them in a loop. Put the earbuds and connector in one hand, and then pull the wire straight out with the other hand. Fold the wire in half and then in half again. You can then put the earphones in your pocket or on a tabletop. By folding the wires, you avoid creating something that looks like a wire ball of Christmas tree lights.

Adding accessories

Beyond earphones, you can find an entire Phone Store full of accessories and baubles that you can obtain for your Android phone. The variety is seemingly endless, and the prices, well, they ain't cheap. Here are some of your choices:

Phone case: Protect your phone by getting it a jacket, one that further expresses your individuality.

Screen protectors: These clear plastic sheets adhere to the touchscreen, protecting it from scratches and other abuse. Ensure that you get screen protectors designed specifically for your phone.

Belt clip: To sate your envy of Batman's utility belt, consider getting a fine leatherette or Naugahyde phone case that you can quickly attach to your belt.

Car charger: This gizmo connects your phone to the car's 12-volt battery source. If you're over 40, the power source was once known as a cigarette lighter.

Car mount: This device holds your Android phone so that you can easily see it while driving. It makes for easier access, although these things are forbidden in some states. See Chapter 17 for information on using Bluetooth in your car for hands-free phone operations.

microSD Card: When your phone supports this type of removable storage, consider buying this memory card. See the earlier section "Installing a microSD card."

Dock: This is a heavy base into which you can set your phone. Some docks are simple cradles that prop up the phone for easy bedside viewing. Others are more sophisticated, offering USB connectors and maybe even a real keyboard.

Wireless charger: Not every phone can be charged wirelessly, but if yours can, definitely get a wireless charger. Set your phone on the pad or prop it up in the dock. The phone's battery starts magically recharging.

Screencasting dongle: This accessory connects to an HDTV or computer monitor. Once configured, it allows you to cast the phone's screen onto the larger-screen device. It's ideal for watching movies, Netflix, or YouTube videos or for enjoying music. Google's Chromecast is an example of a screencasting gizmo.

A Home for Your Phone

Houses built in the 1930s and 1940s often featured a special hole in the wall called a *phone cubby.* It was the shrine in which you would find the home's hard-wired phone. You should treat your Android phone with similar reverence and respect.

Toting your Android phone

The compactness of the modern cell phone makes it perfect for a pocket or even the teensiest of party purses. And its well-thought-out design means that you can carry your phone in your pocket or handbag without fearing that something will accidentally turn it on, dial Mongolia, and run up a heck of a cell phone bill.

Your Android phone most likely features a proximity sensor, so you can even keep the phone in your pocket while you're on a call. The proximity sensor disables the touchscreen, which ensures that nothing accidentally gets pressed when you don't want it to be pressed.

WARNING

Don't forget that you've placed the phone in your pocket, especially in your coat or jacket. You might accidentally sit on the phone, or it can fly out when you peel off your coat. The worst fate for any cell phone is to take a trip through the wash. I'm sure your phone has nightmares about it.

Storing the phone

I recommend that you find a single place for your phone when you're not taking it with you: on top of your desk or workstation, in the kitchen, on the nightstand — you get the idea. Phones are as prone to being misplaced as are your car keys and glasses. Consistency is the key to finding your phone.

Then again, your phone rings, so you can always have someone else call your cell phone to help you locate it.

>> While working, I keep my phone next to my computer. Conveniently, I have the charger plugged into the computer so that the phone remains plugged in, connected, and charging when I'm not using it.

>> Phones on coffee tables get buried under magazines and are often squished when rude people put their feet on the furniture.

>> Avoid putting your phone in direct sunlight; heat is bad news for any electronic gizmo.

Chapter 2

The On/Off Chapter

t would be delightful if your Android phone were smart enough to pop out of the box, say "Hello," and immediately know everything about you. It doesn't, of course. That introduction is still necessary, and it requires some careful attention. It's all part of the initial setup-and-configuration process that happens when you initially turn on the phone. And if you haven't yet turned on your phone, this chapter shows you how.

Hello, Phone

Modern, technical gizmos lack an on–off switch. Instead, they feature a power button. In the case of your Android phone, the button is called the Power/Lock key. This key is used in several ways, some of which may not be obvious or apparent.

TIP

REMEMBER

» The setup process works more smoothly when you already have a Google, or Gmail, account. If you lack such an account, you're prompted to create one in the setup process.

» The phone won't start unless the battery is charged. See Chapter 1.

Turning on your phone for the first time

The very first time you turn on an Android phone is a special occasion. That's when you're required to work through the setup-and-configuration process. Don't worry: It needs to be done only once. After that, turn on your phone according to the directions in the next section.

The specifics for the setup and configuration differ depending on the phone's manufacturer and cellular provider. Odds are pretty good that the people at the Phone Store helped you through the initial setup process. If not, read the generic Android phone setup process outlined in this section, and see the notes at the end of the section for details that may apply to your specific phone.

1. **Press the Power/Lock key to turn on the phone.**

You may have to press and hold the key. When you see the phone's logo on the screen, you can release the key.

It's okay to turn on the phone while it's plugged in and charging.

TIP

2. **Answer the questions presented.**

You're asked to perform some, if not all, of the following activities:

- Select your language

- Activate the phone on the mobile data network

- Choose a Wi-Fi network (can be done later)

- Set the time zone

- Sign in to your Google account

- Add other online accounts

- Set location information

When in doubt, just accept the standard options as presented to you during the setup process. Or you can tap the SKIP button to return to that step later.

Use the onscreen keyboard to fill in text fields. See Chapter 4 for keyboard information.

Other sections in this chapter, as well as throughout this book, offer information and advice on the configuration options and settings. So don't worry if you make a mistake; your selection can be changed later.

3. **After each choice, tap the NEXT button or icon.**

The button might be labeled with the text *NEXT*, or it may appear as the Triangle icon, shown in the margin.

4. **Tap the FINISH button.**

The FINISH button appears on the last screen of the configuration procedure.

From this point on, starting the phone works as described in the next section.

After the initial setup, you see the phone's Home screen. Chapter 3 offers details on using the Home screen, which you should probably read right away, before the temptation to play with your new phone becomes unbearable.

>> You may find yourself asked various questions or prompted to try various tricks as you explore your phone. Some of those prompts are helpful, but it's okay to skip them. To do so, tap the OK I GOT IT button and the SKIP button, and if present, select the DO NOT SHOW AGAIN check box.

>> Additional information on connecting your phone to a Wi-Fi network is found in Chapter 17.

>> Location items relate to how the phone knows its position on Planet Earth. I recommend activating all these items to get the most from your Android phone.

>> It's not necessary to use any specific software provided by the phone's manufacturer or your cellular provider. For example, if you don't want a Samsung account, you don't need to sign up for one; skip that step.

>> Through your Google account, you coordinate your new Android phone with whatever information you already have on the Internet. These details include your Gmail messages, contacts, Google Calendar appointments, photos, music, books, and other Google account details.

>> See the later sidebar "Who is this Android person?" for more information about the Android operating system.

Turning on the phone

To turn on your Android phone, press and hold the Power/Lock key. After a few seconds, you feel the phone vibrate slightly and then see the phone's startup animation, logo, or hypnotic brainwashing image. Release the Power/Lock key; the phone is starting.

Eventually, you see the phone's Unlock screen. See the later section "Working the screen lock" for what to do next.

If you've encrypted your phone's data, you must work a screen lock before the device fully starts. See Chapter 21 for details on encrypting your phone's data.

Unlocking the phone

Most of the time, you don't turn your phone off and on. Instead, you lock and unlock it. To unlock and use the phone, press the Power/Lock key. A quick press is all that's needed. The phone's touchscreen comes to life, and you see one of several types of screen lock. Working these locks is covered in the next section.

After you work the screen lock, you next see the Home screen. Chapter 3 covers how to interact with the Home screen.

>> On a Samsung phone, you can press the Home button to unlock the device. The Home button is centered below the touchscreen.

>> Removing the S Pen on a Samsung Galaxy Note unlocks the phone.

>> The phone lets you answer or decline an incoming call without having to unlock the device. See Chapter 5 for more information on answering, declining, and ignoring incoming calls.

Working the screen lock

When you unlock your phone, you see the lock screen, illustrated on a variety of Android phones in Figure 2-1. The lock screen shows a screen lock, such as the standard Android swipe lock, illustrated in the figure.

FIGURE 2-1: Android phones' lock screen varieties.

To work the Swipe lock, swipe the screen in one direction. Some phones may show a Lock icon, which you drag to unlock the phone. Onscreen animation may assist you with working the Swipe lock.

The Swipe lock isn't a difficult lock to pick. If you've added more security, you might see any one of several different screen locks on your phone. Here are the common Android screen locks:

Swipe: The standard screen lock. Swipe your finger on the screen to unlock the device, as illustrated in Figure 2-1. After swiping, you may see another, more secure screen lock. If so, work it next.

Pattern: Trace a preset pattern over dots on the screen.

PIN: Use the onscreen keyboard to type a number to unlock the device.

Password: Type a password, which can include letters, numbers, and symbols.

None: When this non-lock is selected, the device lacks a screen lock and you can use the phone immediately after pressing the Power/Lock key.

Some phones provide additional types of screen locks. The most common is the fingerprint lock. To unlock the phone, you tap on or slide your finger over the fingerprint-reading gizmo

Other screen locks are available, including the Face Unlock and the Signature lock found on the Samsung Galaxy Note line of phones. For further details on screen locks and how to configure them, see Chapter 21.

Unlocking and running an app

The phone's lock screen may feature app icons. You'll find them at the bottom of the screen (refer to Figure 2-1), if they're available. Use those icons to unlock the phone and run the given app.

For example, to place a quick phone call, swipe the Phone icon up the screen. The phone unlocks and the Phone app appears. Similarly, you can swipe the Camera app icon to snap a quick photo.

>> Some phones let you customize the lock screen apps, such as the Galaxy Note, shown in the center of Figure 2-1.

>> When a secure screen lock is set, the phone isn't actually unlocked when the lock screen app runs. To do anything else with the phone, you must eventually work the screen lock.

WHO IS THIS ANDROID PERSON?

Just like a computer, your phone has an operating system. It's the main program in charge of all the software (apps) inside the phone. Unlike on a computer, however, Android is a mobile device operating system, designed primarily for use in cell phones and tablets.

Android is based on the Linux operating system, which is also a computer operating system, though it's much more stable and bug-free than Windows, so it's not as popular. Google owns, maintains, and develops Android, which is why your online Google information is synced with the phone.

The Android mascot, shown here, often appears on Android apps or hardware. He has no official name, though most folks call him Andy.

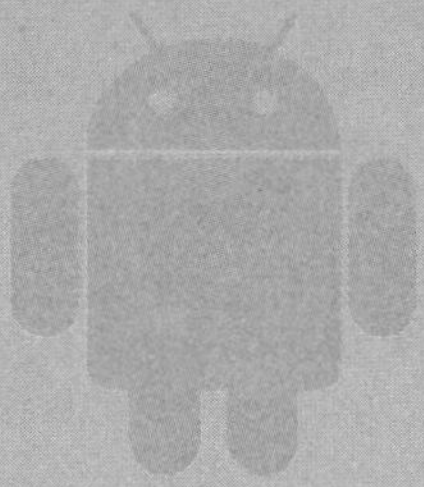

Add More Accounts

Your Android phone can be home to your various online incarnations. This list includes your email accounts, online services, social networking, subscriptions, and other digital personas. I recommend adding those accounts to your phone as you continue the setup-and-configuration process.

With your phone on and unlocked, follow these steps:

1. **Tap the Apps icon.**

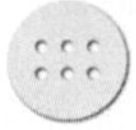

   The Apps icon is found at the bottom of the Home screen. It looks similar to the icon shown in the margin, although on your phone it may look different. See Chapter 3 for the variety.

After tapping the Apps icon, you see the Apps drawer, which lists all apps available on your phone.

2. **Open the Settings app.**

You may have to swipe the Apps drawer screen a few times, paging through the various icons, to find the Settings app.

After you tap the Settings icon, the Settings app runs. You use this app to configure phone options and features.

3. **Choose the Accounts category.**

On some Samsung phones, first tap the General tab atop the Settings app screen to locate the Accounts category.

The category may be titled Accounts and Sync on some phones.

Upon success, you see all existing accounts on your phone, such as email accounts, social networking, cloud storage, and whatever else you may have already set up. If the list is empty, well, it's time to add more accounts!

4. **Tap Add Account.**

You see a list of account types the phone can add for you, such as the list shown in Figure 2-2.

5. **Choose an account from the list.**

For example, to add a Facebook account, choose Facebook from the list.

Don't worry if you don't see the exact type of account you want to add. You may have to install a specific app before an account appears. Chapter 16 covers installing new apps on your phone.

6. **Follow the directions on the screen to sign in to your account.**

The steps that follow depend on the account. Generally speaking, you sign in using an existing username and password.

 You can continue adding accounts by repeating these steps. When you're done, tap the Home navigation icon to return to the Home screen.

>> See Chapter 9 for specifics on adding email accounts to your Android phone.

>> Chapter 11 covers social networking on your phone and offers advice on adding those types of accounts.

FIGURE 2-2:
Common
account types
on an Android
phone.

Navigation icons

Back   Home   Recent

TRANSFERRING INFORMATION FROM YOUR OLD PHONE

Here's one task you don't need to worry about: All the Google information associated with your old phone — or any other Android device, including a tablet — is instantly transferred to your new phone. This information includes contacts, Gmail, events, and other Googly account data. You can even install apps you've previously obtained (free or purchased).

As you add accounts to your phone, the information associated with those accounts is migrated to the device. You might also see media, such as photos, videos, and music transferred, if you use online sharing services to host the media. For information that's not transferred, see Chapter 18, which covers methods for moving files between your phone and other devices.

Goodbye, Phone

You can dismiss your Android phone in several ways, only two of which involve using a steamroller or raging elephant. The other methods are covered in this section.

Locking the phone

To lock your Android phone, press and release the Power/Lock key. The touch-screen display turns off and the phone is locked.

REMEMBER

>> Some phones may not turn off the display while they're locked. You may see the current time and notifications displayed, albeit on a very dim screen. This feature doesn't affect the phone's battery life and in many cases this setting can be changed if you don't like the always-on touchscreen. See Chapter 20 for details.

>> Your phone will spend most of its time locked. The phone still works while locked; it receives email and plays music and signals alerts and alarms. Phone calls arrive. Yet, while the phone is locked, it doesn't use as much power as it does when the display is on.

>> You can lock the phone while you're making a call. Simply press and release the Power/Lock key. The call stays connected, but the display is disabled.

>> Locking doesn't turn off the phone.

>> Refer to Chapter 20 for information on setting the automatic lock time-out value.

Turning off the phone

To turn off your phone, obey these steps:

1. **Press and hold down the Power/Lock key.**

   When you see the Phone Options card, similar to what's shown in Figure 2-3, you can release the key.

2. **Tap the Power Off item.**

   If a confirmation message appears, tap the OK button. The phone turns itself off.

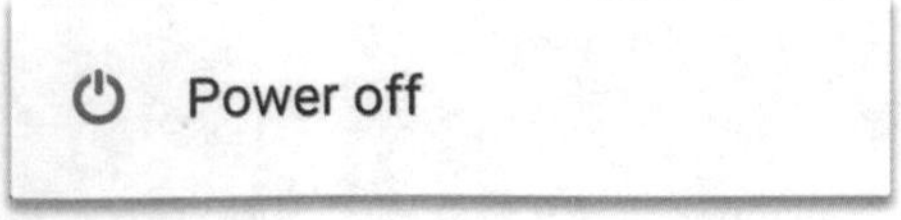

FIGURE 2-3:
The Phone
Options card.

⏻   **Power off**

The phone doesn't receive calls when it's turned off. Likewise, any alarms or reminders you've set won't sound when the phone is off.

» Calls received while the phone is off are routed instead to voicemail.

» The phone can be charged while it's off.

» The Phone Options card might sport more options than what are shown in Figure 2-3. The bare minimum is the Power Off item. Other items might include Restart, Sleep, Kid Mode, as well as volume and vibration settings.

Chapter 3

The Android Tour

I t used to be that you could judge a computer's power based on its number of buttons, knobs, and dials. Those room-size computers back in the 1960s had hundreds of switches and blinking lights. So it makes sense that super-sophisticated computers of the future would have even more buttons, knobs, and blinking lights. The reality is far different.

Today, devices like your Android phone are basically bereft of buttons. Despite its seemingly endless potential, your phone features a touchscreen as its main control. That's probably not what folks expected back in the 1960s, and it's probably not what you expect today, especially if you anticipate a button, knob, or dial for every phone function.

Basic Operations

Your Android phone's capability to frustrate you is only as powerful as your fear of the touchscreen and how it works. After you clear that hurdle, understanding how your phone works becomes easier.

Manipulating the touchscreen

The touchscreen works in combination with one or two of your fingers. You can choose which fingers to use, or whether to be adventurous and try using the tip of your nose, but touch the screen you must. Here are some of the many ways you manipulate your phone's touchscreen:

Tap: In this simple operation, you touch the screen. Generally, you're touching an object such as an icon or graphical control. A tap might also be called *touch* or *press.*

Double-tap: Touch the screen twice in the same location. A double-tap can be used to zoom in on an image or a map, but it can also zoom out. Because of the double-tap's dual nature, I recommend using the pinch and spread operations instead.

Long-press: Tap part of the screen and hold down your finger. Depending on what you're doing, a pop-up menu may appear, or the item you're long-pressing may get "picked up" so that you can drag (move) it around. Long-press might also be referred to as tap and hold.

Swipe: To swipe, tap your finger on one spot and then move your finger to another spot. Swipes can go up, down, left, or right. This action often moves information on the touchscreen, similar to the way scrolling works on a computer. A swipe can be fast or slow. It's also called a *flick* or *slide.*

Drag: A combination of long-press and then swipe, the drag operation moves items on the screen. Start with the long press, and then keep your finger on the screen to swipe. Lift your finger to complete the action.

Pinch: A pinch involves two fingers, which start out separated and then are brought together. The pinch is used to zoom out on an image or a map. This move may also be called a *pinch close.*

Spread: In the opposite of a pinch, you start with your fingers together and then spread them. The spread is used to zoom in. It's also known as a *pinch open.*

Rotate: Use two fingers to twist around a central point on the touchscreen, which has the effect of rotating an object on the screen. If you have trouble with this operation, pretend that you're turning the dial on a safe.

REMEMBER

You can't manipulate the touchscreen while wearing gloves, unless they're gloves specially designed for use with electronic touchscreens, such as the gloves that Batman wears.

Selecting a group of items

Unique to your phone's touchscreen is the manner in which a group of items is selected. On a computer, you drag the mouse over the items. On a phone's touchscreen, you perform these steps:

1. Long-press the first item, such as a photo thumbnail in an album.

The item is selected, and it appears highlighted on the screen or grows a tiny check mark. Also, an action bar appears atop the screen, similar to the one shown in Figure 3-1. It lists icons such as Share and Delete, which manipulate the group of selected items.

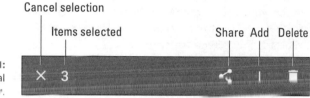

Cancel selection

Items selected

Share Add Delete

FIGURE 3-1:
A typical
action bar.

2. Tap additional items to select them.

As long as the action bar appears, you tap to keep adding items to the mix. The action bar lists the total number of selected items, as illustrated in Figure 3-1.

3. Do something with the group.

Choose an icon from the action bar.

To cancel the selection, tap the Cancel (X) icon on the action bar, which deselects all items.

Using the navigation icons

Below the touchscreen dwells a series of icons. They can appear as part of the touchscreen itself, or on some phones they may be part of the bezel or even be physical buttons. These are the navigation icons, and they serve specific and uniform functions throughout the Android operating system.

Traditionally, you find three navigation icons: Back, Home, and Recent. Table 3-1 illustrates how the navigation icons look for the most recent releases of the Android operating system.

TABLE 3-1	**Navigation Icon Varieties**	
Icon	Lollipop, Marshmallow, Nougat	Kit Kat and Earlier
Home	◯	⌂
Back	◁	↩
Recent	☐	⧉

Home: No matter what you're doing on the phone, touch this icon to display the Home screen. When you're already viewing the Home screen, tap this icon to view the main or center Home screen page.

Back: The Back icon serves several purposes, all of which fit neatly under the concept of "back." Tap the icon once to return to a previous page, dismiss an onscreen menu, close a card, and so on.

The Back navigation icon may change its orientation, as shown in the margin. Tap this icon to hide the onscreen keyboard, dismiss dictation, or perform other actions similar to using the Back navigation icon.

Recent: Tap the Recent icon to see the *Overview,* a list of recently opened or currently running apps. See the later section "Switching between running apps" for more info on the Overview.

The three navigation icons may hide themselves when certain apps run. To access the icons, tap the screen. For some full-screen apps and games, swipe the screen from top to bottom to see the navigation icons.

Setting the volume

The Volume key is located on the side of the phone. Press the top part of the key to raise the volume. Press the bottom part to lower the volume.

As you press the Volume key, a card appears on the touchscreen to illustrate the relative volume level, as shown in Figure 3-2.

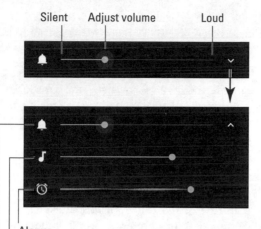

Silent   Adjust volume   Loud

Alarms

Media

General

Not every volume card looks like the one shown in Figure 3-2, though they all feature a slider control. Additional controls let you set specific volumes; tap an icon on the card to view details. If the Settings icon appears on the card, use it to make more specific adjustments.

TIP

>> The Volume key controls whatever noise the phone is making when you use it: When you're on a call, the volume controls set the call level. When you're listening to music or watching a video, the volume controls adjust those sounds.

>> You can set the volume when the phone is locked. That means you don't need to unlock the phone to adjust the volume while you're listening to music.

>> See Chapter 20 for more volume control information.

"Silence your phone!"

How many times have you heard the admonition "Please silence your cell phones"? The quick way to obey this command with your Android phone is to unlock the phone and keep pressing the bottom part of the Volume key until the phone vibrates. You're good to go.

TIP

>> Some phones feature a Mute action on the Phone Options card: Press and hold the Power/Lock key and then choose Mute or Vibrate.

>> You might also find a Sound quick setting. Tap that icon to mute or vibrate the phone. See the later section "Accessing quick settings" for details.

>> When the phone is silenced or in Vibration mode, an appropriate status icon appears on the status bar. The stock Android status icon is shown in the margin.

>> You make the phone noisy again by reversing the directions in this section. Most commonly, press the "louder" end of the Volume key to restore the phone's sound.

Changing the orientation

Android phones feature a gizmo called an *accelerometer*. Various apps use it to determine in which direction the phone is pointed or whether you've changed the phone from an upright to a horizontal position.

To demonstrate how the phone orients itself, rotate the device to the left or right. Most apps change their presentation between vertical and horizontal to match the phone's orientation. (The web browser app is a good app to use for testing the orientation.)

>> The orientation feature may not work for all apps — specifically, games that present themselves in one format only. Also, some phones' Home screen may not reorient.

>> The onscreen keyboard is more usable when the phone is in its horizontal orientation. See Chapter 4.

>> Some phones feature a quick setting that locks the orientation. See the later section "Accessing quick settings."

>> A nifty app that demonstrates the phone's accelerometer in action is the game Labyrinth. You can purchase it at Google Play or download the free version, Labyrinth Lite. See Chapter 16 for information on Google Play.

Home Screen Chores

The *Home screen* is where you start your Android day. It's the location from which you begin common phone duties, such as starting an app. Knowing about the Home screen is an important part of understanding your Android phone.

To view the Home screen at any time, tap the Home navigation icon found at the bottom of the touchscreen. Some phones feature a physical Home button, which performs the same duties as the Home navigation icon.

Exploring the Home screen

A typical Android Home screen is illustrated in Figure 3-3. Several fun and interesting things appear on the Home screen. Discover these items on your own phone's Home screen:

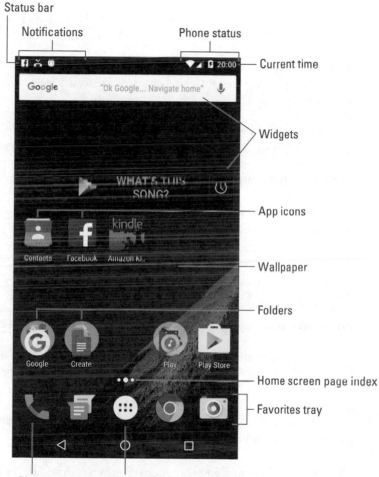

FIGURE 3-3:
The Home
screen.

Status bar: The top of the Home screen is the status bar. It contains notification icons and status icons, plus the current time. If the status bar disappears, a quick swipe from the top of the screen downward redisplays it.

Notifications: These icons come and go, depending on what happens in your digital life. For example, a new notification icon appears whenever you receive a new email message or have a pending appointment. See the later section "Reviewing notifications."

Phone status: Icons on the right end of the status bar represent the phone's current condition, such as the type of network connection, signal strength, Wi-Fi status, battery charge, and other items.

App launchers: These icons represent apps installed on your phone. Tap a launcher to run the associated app.

Widgets: Widgets work like teensy programs that display information or let you control the phone, manipulate a feature, access an app, or do something purely amusing.

Folders: Multiple apps can be stored in a folder. Tap the folder to open it and view the app launchers stored within. See Chapter 19 for more information on folders.

Wallpaper: The background image you see on the Home screen is the wallpaper.

Home screen page index: This series of dots helps you navigate between the various Home screen pages. See the next section.

Favorites tray: The lineup of launcher icons near the bottom of the screen is the favorites tray. It shows the same launchers at the bottom of every Home screen page.

Phone app: You use the Phone app to make calls. It's kind of a big deal.

Apps icon: Tap this icon to view the Apps drawer, a collection of all apps available on your phone. See the later section "Finding an app in the Apps drawer."

REMEMBER

Ensure that you recognize the names of the various parts of the Home screen. These terms are used throughout this book and in whatever other scant Android phone documentation exists.

>> The Home screen is entirely customizable. You can add and remove icons, add widgets and shortcuts, and even change wallpaper images. See Chapter 20 for more information.

>> Touching part of the Home screen that doesn't feature an icon or a control does nothing — unless you're using the live wallpaper feature. In that case, touching the screen changes the wallpaper in some way, depending on the wallpaper that's selected. You can read more about live wallpaper in Chapter 20.

>> You may see numbers affixed to certain launcher icons. Those numbers indicate pending actions, such as unread email messages, indicated by the icon shown in the margin.

Switching Home screen pages

The Home screen is more than what you see. It's actually an entire street of Home screens, with only one Home screen page visible at a time.

To switch from one page to another, swipe the Home screen left or right. Use the Home screen page index (refer to Figure 3-3) to determine which Home screen page you're viewing.

>> On some phones, the main Home screen page is shown by a House icon on the Home screen page index.

>> See Chapter 20 for information on managing Home screen pages, adding new pages, and removing pages.

>> The far left Home screen page on some phones is the Google Now app. See Chapter 15 for information on Google Now.

Reviewing notifications

Notifications appear as icons at the top left of the Home screen, as illustrated earlier, in Figure 3-3. To review them, pull down the notifications drawer by dragging your finger from the top of the screen downward. Two variations on the notifications drawer presentation are illustrated in Figure 3-4.

Peruse the notifications by swiping them up and down. To deal with a specific notification, tap it. What happens next depends on the notification, although typically the app that generated the notification runs and shows more details.

After a notification is chosen, it disappears. Or you can dismiss a notification by swiping it to the right or left. To dismiss all notifications, tap the Clear icon, as shown in the margin, or the CLEAR button. If the Clear icon isn't visible, wipe the notifications drawer up or down to locate it.

To hide the notifications drawer, swipe the notifications drawer upward on the screen. Or, if you find this process frustrating (and it can be), tap the Back navigation icon.

Quick Settings

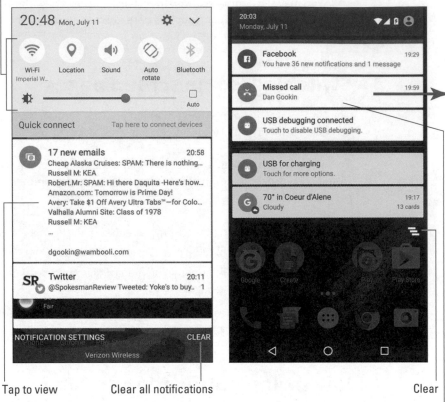

FIGURE 3-4:
Reviewing
notifications.

Tap to view   Clear all notifications   Clear

Swipe to dismiss

>> Notifications can stack up if you don't deal with them!

>> When more notifications are present than can be shown, the More Notifications icon appears on the far left end of the status bar, similar to what's shown in the margin.

>> Some ongoing notifications cannot be dismissed. For example, the USB Connection notification stays active until the USB cable is disconnected.

>> Swiping away some notifications doesn't prevent them from appearing again in the future. For example, notifications to update apps continue to appear, as do calendar reminders.

>> Some apps, such as Facebook and Twitter, don't display notifications unless you're logged in. See Chapter 11.

>> New notifications are heralded by a notification ringtone. Chapter 20 provides information on changing the sound.

>> Notifications may also appear on the phone's lock screen. Controlling which types of notifications appear is covered in Chapter 21.

Accessing quick settings

The quick settings appear as large buttons or icons atop the notifications drawer. Use them to access popular phone features or turn settings on or off, such as Bluetooth, Wi-Fi, Airplane Mode, Auto Rotate, and more.

On some phones, such as the Samsung phone shown on the left in Figure 3-4, the quick settings appear all the time atop the navigation drawer. Swipe the Quick Settings icons left or right to view the lot. Other phones require that you swipe the notifications drawer downward twice to see quick settings, as illustrated in Figure 3-5.

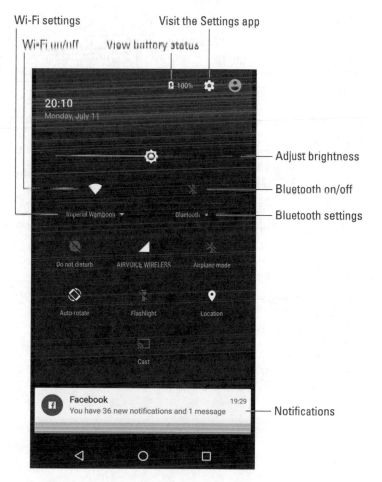

FIGURE 3-5:
Quick Settings drawer.

Tap the Back or Home navigation icons to dismiss the Quick Settings drawer. If you have excellent touchscreen skills, swipe the screen from bottom to top to banish the Quick Settings drawer.

The World of Apps

The Android operating system can pack thrill-a-minute excitement, but it's probably not the only reason you purchased the phone. No, an Android phone's success lies with the apps you obtain. Knowing how to deal with apps is vital to becoming a successful, happy Android phone user.

Starting an app

To start an app, tap its icon. The app starts.

Apps can be started from the Home screen: Tap a launcher icon to start the associated app. Apps can be started also from the Apps drawer, as described in the later section "Finding an app in the Apps drawer."

You can also start an app found in a Home screen folder: Tap to open the folder. Tap to open the app.

Quitting an app

Unlike on a computer, you don't need to quit apps on your Android phone. To leave an app, tap the Home navigation icon to return to the Home screen. You can keep tapping the Back navigation icon to back out of an app. Or you can tap the Recent navigation icon to switch to another running app.

TECHNICAL STUFF

>> Occasionally, you'll find an app that features a Quit command or an Exit command, but for the most part, you don't quit an app on your phone like you quit a program on a computer.

>> If necessary, the Android operating system shuts down apps you haven't used in a while. You can directly stop apps that have run amok, as described in Chapter 19.

WONDERFUL WIDGETS

Like apps, widgets appear on the Home screen. To use a widget, tap it. What happens after that depends on the widget and what it does.

For example, the YouTube widget lets you peruse videos. The Calendar widget shows a preview of your upcoming schedule. A Twitter widget may display recent tweets. Other widgets do interesting things, display useful information, or give you access to the phone's settings or features.

New widgets are obtained from Google Play, just like apps. See Chapter 16 for information. Also see Chapter 19 for details on adding widgets to the Home screen.

Finding an app in the Apps drawer

The place where you find all apps installed on your Android phone is the Apps drawer. To access the Apps drawer, tap the Apps icon on the Home screen. This icon has a different look to it, depending on your Android phone. Figure 3-6 illustrates several varieties of the Apps icon.

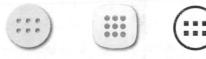

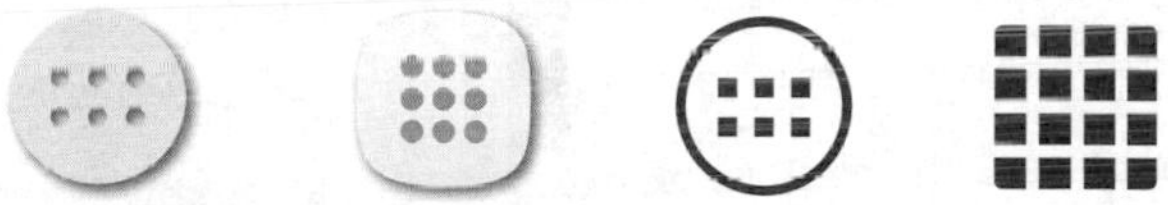

FIGURE 3-6: Apps icon varieties.

After you tap the Apps icon, you see the Apps drawer. Swipe through the pages left and right or up and down across the touchscreen.

To run an app, tap its icon. The app starts, taking over the screen and doing whatever magical thing the app does.

TIP

- As you add new apps to your phone, they appear in the Apps drawer. See Chapter 16 for information on adding new apps.
- The Apps drawer may list frequently opened apps at the top of the list. These apps are good candidates for adding to the Home screen. See Chapter 19 for details on adding app launchers to the Home screen.
- Some phones allow you to create folders in the Apps drawer. These folders contain multiple apps, which helps keep things organized. To access apps in the folder, tap the Folder icon.

» The Apps drawer displays apps alphabetically. On some phones, you can switch to a non-alphabetical viewing grid. With that feature active, it's possible to rearrange the apps in any order you like.

Switching between running apps

The apps you run on your phone don't quit when you dismiss them from the screen. For the most part, they stay running. To switch between running apps, or any app you've recently opened, tap the Recent navigation icon. You see the Overview, similar to what's shown in Figure 3-7.

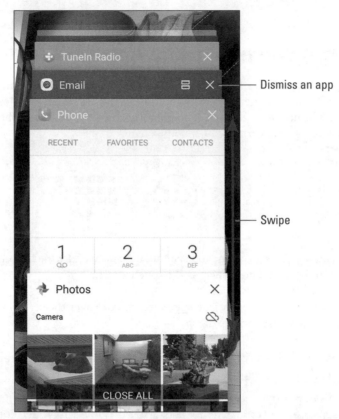

Dismiss an app

Swipe

FIGURE 3-7:
The Overview shows recently opened apps.

Swipe the list to view all the apps. Tap the app's card to switch to that app. To exit from the Overview, tap the Back navigation icon.

» You can remove an app from the Overview by swiping it off the list or tapping the X (close) icon, as illustrated in Figure 3-7.

>> Removing an app from the Overview is pretty much the same thing as quitting an app.

>> For Android phones without a Recent icon, long-press the Home navigation icon to see the Overview.

REMEMBER

>> The Android operating system may shut down apps that haven't received attention for a while. Don't be surprised when you see an app missing from the Overview. If so, just start it up again as you normally would.

Common Android Icons

In additional to the navigation icons, the Android operating system features a consistent armada of other, helpful icons. These icons serve common and consistent functions in apps as well as in the Android operating system. Table 3-2 lists the most common of these icons and their functions.

TABLE 3-2   **Common Icons**

Icon	Name	What It Does
	Action Bar	Displays a pop-up menu. This teensy icon appears in the lower right corner of a button or an image, indicating that actions (commands) are attached.
	Action Overflow	Displays a list of actions, similar to a menu.
	Add	Adds or creates an item. The plus symbol (+) may be used in combination with other symbols, depending on the app.
	Close	Dismisses a card or clears text from an input field.
	Delete	Removes one or more items from a list or deletes a message.
	Dictation	Lets you use your voice to dictate text.
	Done	Dismisses an action bar, such as the text-editing action bar.

(continued)

TABLE 3-2 *(continued)*

Icon	Name	What It Does
	Edit	Lets you edit an item, add text, or fill in fields.
	Favorite	Flags a favorite item, such as a contact or a web page.
	Refresh	Fetches new information or reloads.
	Search	Searches the phone or the Internet for a tidbit of information.
	Settings	Adjusts options for an app.
	Share	Shares information stored on the phone via email, social networking, or other Internet services.

Various sections throughout this book give examples of using the icons. Their images appear in the book's margins where relevant.

» The Share icon shown in Table 3-2 has an evil twin, shown in the margin. Both icons represent the Share action.

» Other common symbols are used as icons in various apps. For example, the standard Play and Pause icons are used in many apps. The Stop icon looks like a small square.

TIP

» If the Refresh icon isn't visible, try "tugging" the screen: Swipe downward starting just below the status bar toward the center of the touchscreen. This action frequently updates or refreshes the app's contents.

» The Action Overflow might also be referred to as *Overflow*.

» Samsung phones may display a MORE button instead of the Action Overflow icon. The MORE button serves the same function.

Chapter 4

Text to Type, Text to Edit

Although I seriously doubt that anyone would consider writing the Great American Novel on an Android phone, you will do a lot of typing on your phone. These tasks include crafting messages, composing email, and taking notes. To help with those text-creation duties, your phone features an onscreen keyboard, and it provides for voice dictation. You can edit the text you create, copy, paste, and all that jazz.

Onscreen Keyboard Mania

The onscreen keyboard appears on the bottom part of the touchscreen whenever your phone demands text as input. Figure 4-1 illustrates the typical Android keyboard, which is called the *Google keyboard*. Your phone may use the same keyboard or a variation that looks subtly different. All onscreen keyboards are based on the traditional QWERTY layout.

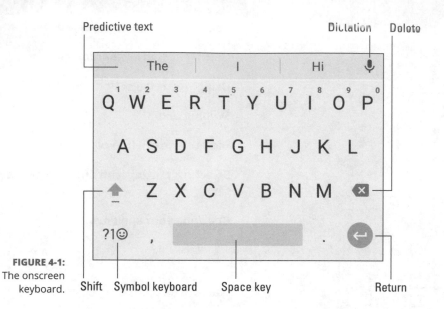

Predictive text   Dictation   Delete

FIGURE 4-1:
The onscreen
keyboard.

Shift   Symbol keyboard   Space key   Return

In Figure 4-1, the onscreen keyboard is shown in alphabetic mode. You see keys from A through Z in uppercase. (Uppercase is set automatically at the start of a line of text). Also visible are the Shift key to produce capital letters and the Delete key to backspace and erase.

The Return key changes its look and function depending on what you're typing. Your keyboard may show these variations graphically or by labeling the key with text. The stock Android symbols are illustrated in Figure 4-2. Here's what each one does:

FIGURE 4-2:
Return key
variations.

Return   Search   Go   Next   Done

Return: Just like the Return or Enter key on a computer keyboard, this key ends a paragraph of text. It's used mostly when filling in long stretches of text or when multiline input is needed.

Search: You see the Search key when you're searching for something. Tap the key to start the search.

Go: This action key directs the app to proceed with a request, accept input, or perform another action.

Next: This key appears when you're typing information into multiple fields. Tap this key to switch from one field to the next, such as when typing a username and password.

Done: This key appears whenever you've finished typing text in the final field and are ready to submit input.

The large key at the bottom center of the onscreen keyboard is the Space key. The keys to the left and right may change depending on the context of what you're typing. For example, a slash (/) key or .com (dot-com) key may appear in order to assist in typing a web page or email address. Other keys may change as well, although the basic alphabetic keys remain the same.

>> To display the onscreen keyboard, tap any text field or spot on the screen where typing is permitted.

>> To dismiss the onscreen keyboard, tap the Back icon. This icon may appear as shown in the margin or as a down-pointing chevron.

>> Some onscreen keyboards feature a multifunction key. It may be labeled with the Settings (Gear) icon, the Microphone icon, an emoji, or some other icon. Long-press the multifunction key to view its options.

TIP

>> The keyboard reorients itself when you rotate the phone. The onscreen keyboard's horizontal orientation is wide, so you might find it easier to use.

>> I haven't seen one in a while, but some older Android phones featured a physical keyboard. If you'd like to use this type of keyboard with your phone, consider getting a Bluetooth keyboard. The models available for use on tablets also work on Android phones. See Chapter 17 for details on Bluetooth.

The Ol' Hunt-and-Peck

Trust me: No one touch-types on a cell phone. No one. Not even those preteens who seem to write text messages at the speed of light. So don't feel bad if you can't type on your Android phone as fast as you can on a computer. On the phone, everything is hunt-and-peck.

Typing one character at a time

The onscreen keyboard is pretty easy to figure out: Tap a letter to produce the character. As you type, the key you touch is highlighted. The phone may give a wee bit of feedback in the form of a faint click or vibration.

>> To type in all caps, tap the Shift key twice. The Shift key may appear highlighted, the shift symbol may change color, or a colored dot may appear on

the key, all of which indicate that Shift Lock is on. Tap the Shift key again to turn off Shift Lock.

>> Above all, it helps to type slowly until you get used to the onscreen keyboard.

>> When you make a mistake, tap the Delete key to back up and erase.

>> A blinking cursor on the touchscreen shows where new text appears, which is similar to how typing text works on a computer.

>> When you type a password, the character you type appears briefly, but for security reasons it's then replaced by a black dot.

REMEMBER

>> People generally accept the concept that composing text on a phone isn't perfect. Don't sweat it if you make a few mistakes as you type text messages or email, though you should expect some curious replies about unintended typos.

>> See the later section "Text Editing," for more details on editing your text.

Accessing special characters

You're not limited to typing only the symbols you see on the alphabetic keyboard. Most Android phones feature alternative character keyboards. To access these special keyboards, tap the symbol or numeric key, such as the ?1☺ key. (Refer to Figure 4-1.)

The number and variety of special character keyboards varies from phone to phone. At least one symbol keyboard is available, though you may find multiple symbol keyboards, special numeric keypads, and even emoji keyboards. In Figure 4-3, four different symbol keyboards are shown.

To switch keyboards, locate the special symbols, illustrated in Figure 4-3. Tap a symbol to view an alternative character set. Though these symbols and the keyboards vary, nearly all Android phones use the ABC key to return to the standard, alpha keyboard.

TIP

Some special symbols are available quickly from the alpha keyboard. These symbols include accented letters and other common characters. The secret is to long-press a key, such as the A key, shown in Figure 4-4.

After you long-press, drag your finger upward to choose a character from the pop-up palette. If you choose the wrong character, tap the Delete key on the onscreen keyboard to erase the mistyped symbol.

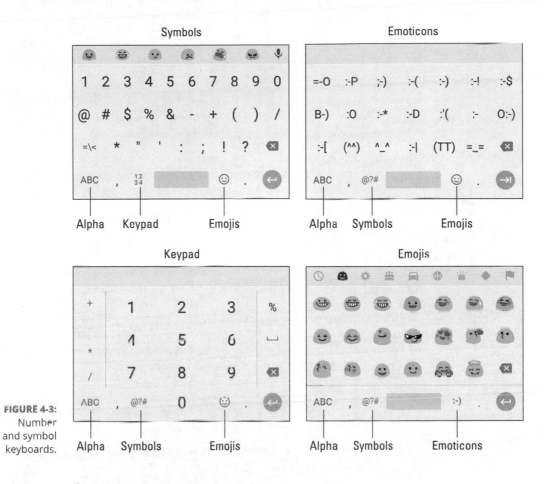

Symbols

Emoticons

FIGURE 4-3:
Number
and symbol
keyboards.

Keypad

Emojis

Alpha   Keypad   Emojis

Alpha   Symbols   Emojis

Alpha   Symbols   Emojis

Alpha   Symbols   Emoticons

Drag your finger over a character to insert

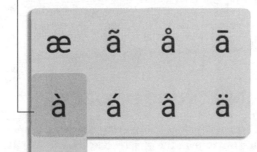

FIGURE 4-4:
Special symbol
pop-up palette
thing.

Long-press

>> Some onscreen keyboards show tiny symbols next to the letters and numbers. If so, you can long-press that key to access the symbol, similar to the technique illustrated in Figure 4-4.

>> Emojis are tiny images that express words or concepts, such as those shown on the bottom right in Figure 4-3. The variety of emojis is almost endless; choose another palette from the top of the keyboard.

>> Emoticons predate emojis. They use characters to create faces or other expressions. Unlike emojis, emoticons appear identical on every smartphone; emojis may look subtly different and not every phone sports the same variety.

TECHNICAL
STUFF

>> Emoticon is a portmanteau of *emotion* and *icon*. *Emoji* is a Japanese term that means "picture character."

Using predictive text to type quickly

As you type, you may see a selection of word suggestions just above the onscreen keyboard. That's the predictive-text feature in action. You can use this feature to greatly accelerate your typing: As you type, tap a word suggestion atop the onscreen keyboard. That word is inserted into the text.

If the desired word doesn't appear, continue typing: The predictive-text feature makes suggestions based on what you've typed so far.

>> On some phones, you may see one of the predictive-text words shown in bold. If so, tap the Space key to insert that word.

>> If three dots appear beneath a predictive-text word, long-press that word to view other word choices.

>> The predictive-text feature is part of the Google Keyboard. If it's not active, see Chapter 20 for directions.

Typing without lifting a finger

If you're really after speed, consider using gesture typing. It allows you to "type" words by swiping your finger over the onscreen keyboard, like mad scribbling but with a positive result.

To use gesture typing, drag a finger over letters on the onscreen keyboard. Figure 4-5 illustrates how the word *hello* is typed in this manner.

Start swiping here

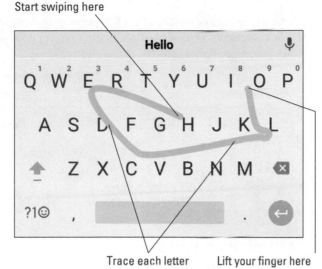

Hello

FIGURE 4-5:
Using gesture
typing to
type *hello*.

Trace each letter   Lift your finger here

Gesture typing is disabled when typing a password or an email address or for other specific activities. When it's unavailable, type one letter at a time.

If gesture typing is unavailable, see Chapter 20 for activation instructions.

Google Voice Typing

Your Android phone has the amazing capability to interpret your utterances as text. It works almost as well as computer dictation in science fiction movies, though I can't seem to find the command to destroy Alderaan.

» Dictation is available whenever you see the Microphone icon. This icon appears on the keyboard as well as in other locations, such as search boxes.

» On some keyboards, the Microphone icon appears on a multifunction key. Long-press that key to choose its dictation function.

» See Chapter 20 for information on enabling dictation, should you have trouble with this feature.

Dictating text

Talking to your phone really works, and works quite well, providing that you tap the Dictation key on the onscreen keyboard and properly dictate your text.

After you tap the Dictation key, a special card appears at the bottom of the screen, similar to the one shown in Figure 4-6. When the text *Tap to Speak* or *Speak Now* appears, dictate your text; speak directly at the phone.

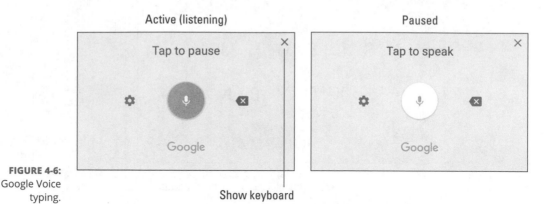

Active (listening)   Paused

Tap to pause   Tap to speak

Google   Google

Show keyboard

FIGURE 4-6:
Google Voice
typing.

As you speak, the Microphone icon on the screen flashes. The flashing doesn't mean that the phone is embarrassed at what you're saying. No, the flashing merely indicates that your words are being digested.

The text you utter appears as you speak. To pause, touch the *Tap to Pause* text on the screen. To resume, tap the *Tap to Speak* text. To dismiss dictation and summon the onscreen keyboard, tap the Cancel (X) icon in the upper right corner of the card. (Refer to Figure 4-6.)

TIP

>> You might see a description the first time you try voice input. Tap the OK button to continue.

>> If you don't like a word that's chosen by the dictation feature, tap the word on the screen. You see a pop-up list of alternatives from which to choose.

>> You can't use Google Voice typing to edit text. Text editing still takes place on the touchscreen, as described in the later section "Text Editing."

>> Speak the punctuation in your text. For example, you would say, "I'm sorry comma and it won't happen again" to produce the text *I'm sorry, and it won't happen again* or something close to that.

>> Common punctuation you can dictate includes: *comma, period, exclamation point, question mark, colon,* and *new line.*

>> You cannot dictate capital letters. If you're a stickler for such things, you'll have to go back and edit the text.

>> Dictation may not work without an Internet connection.

Uttering s**** words

WARNING

Your Android phone features a voice censor. It edits those naughty words you might utter, keeping the word's first letter and replacing the rest with an appropriate number of asterisks.

For example, if *spatula* were a blue word and you uttered "spatula" when dictating text, the dictation feature would place s****** on the screen rather than the word *spatula*.

Yeah, I know: silly. Or "s****."

>> Your Android phone knows a lot of blue terms, including the infamous "Seven Words You Can Never Say on Television," but apparently the terms *crap* and *damn* are fine. Don't ask me how much time I spent researching this topic.

>> See Chapter 24 if you feel the need to disable this feature.

Text Editing

You probably won't do a lot of text editing on your Android phone. Well, no major editing, such as for a term paper or ransom note. From time to time, however, you may find yourself wanting to fix a word. It's usually a sign that you're over 25; kids no longer seem to care about editing text.

Moving the cursor

The first part of editing text is to move the cursor to the right spot. The cursor is that blinking, vertical line where text appears. From that point, you can type, edit, or paste or simply marvel at the blinking cursor.

On a computer, you use the keyboard or a pointing device to move the cursor. On your Android phone, you use your finger: Tap the spot in the text where you want the cursor to appear. To help your accuracy, a cursor tab appears below the text, similar to the one shown in the margin. You can drag that tab to precisely locate the cursor in the text.

After you move the cursor, you can continue to type, use the Delete key to back up and erase, or paste text copied from elsewhere.

You may see a button appear above the cursor tab containing a Paste action. Use that button to paste text, as described in the later section "Cutting, copying, and pasting text."

Selecting text

Selecting text on an Android phone works just like selecting text in a word processor: You mark the start and end of a block. That chunk of text appears highlighted on the screen. How you get there, however, can be a mystery — until now!

To select text, long-press a word. Upon success, you see a chunk of text selected, as shown in Figure 4-7.

Drag the start and end markers around the touchscreen to define the block of selected text. Use the action bar that appears in order to choose what do to with the text, as shown in Figure 4-7. Tap the action overflow to view additional commands.

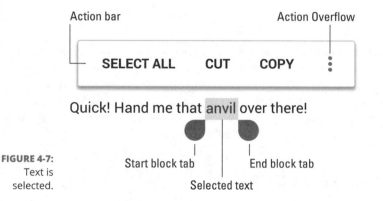

FIGURE 4-7:
Text is
selected.

To cancel text selection, tap elsewhere in the text.

>> On some phones, the action bar may appear at the top of the screen when text is selected. Icons might be used instead of the text buttons illustrated in Figure 4-7.

>> See the next section for information on cutting and copying text. To delete a selected block of text, tap the Delete key on the onscreen keyboard. To replace the text, type something new. To replace the text with previously cut or copied text, tap the Action Overflow and choose the PASTE action.

>> Selecting text on a web page works the same as selecting text in any other app. The difference is that text can only be copied from the web page, not cut or deleted.

>> Seeing the onscreen keyboard is a good indication that you can edit and select text.

>> The action bar's Select All command marks all text as a single block.

Cutting, copying, and pasting text

Selected text is primed for cutting or copying, which works on your phone just like it does in your favorite word processor. After you select the text, choose the proper command from the action bar: COPY to copy the text or CUT to cut the text.

Text that is cut or copied on your phone is stored in a clipboard. To paste any previously cut or copied text, position the cursor on the spot where you want the text pasted. A PASTE button appears above the cursor tab, as shown in the margin. Tap the button to paste the text.

>> Some phones feature a Clipboard app, which lets you peruse, review, and select previously cut or copied text or images. You might even find the Clipboard icon on the action bar or onscreen keyboard.

>> You can paste text only into locations where text is allowed. Odds are good that if you see the onscreen keyboard, you can paste text.

Dealing with speling errors

Similar to a word processor, your Android phone may highlight misspelled words, presenting them with an intimidating red underline.

To remedy the situation, tap the red-underlined word. Choose a replacement word from the pop-up list of alternatives. If the word is correctly spelled but unknown to the phone, choose to add the word to a personal dictionary.

>> Most words are autocorrected as you type them. To undo an autocorrection, tap the word again. Choose a replacement word from the predictive-text list, or tap the REPLACE button to see more options.

>> Yes! Your phone has a personal dictionary. See Chapter 24 for details.

2

Reach Out and Touch Someone

IN THIS PART . . .

Understand how to make phone calls.

Explore voicemail and Google Voice.

Use the address book.

Try text messaging.

Send and receive email.

Discover the web on a cell phone.

Connect with social networking.

Chapter 5
It's a Telephone

The patent for the telephone was awarded to Alexander Graham Bell in 1876. Telephone-like devices existed before then, and a host of 19th century scientists worked on the concept. Bell beat them all to the patent office, so he gets the credit. And it took many more years for people to patent other, ancillary inventions, including the busy signal (1878), the notion of a second line for teenage girls (1896), and the extension cord, which allowed for simultaneously talking and pacing (1902).

Making a phone call is simple enough; it's the reason you have an Android phone and not an Android tablet. Phone calls aren't the limit of what your device can do. Features such as call waiting, speed dial, call forwarding, and multiple calls are also possible. These features and more are presented in this chapter.

Reach Out and Touch Someone

Until they perfect teleportation, making a phone call is truly the next best thing to being there. It all starts by punching in a number or calling someone from the phone's address book.

Placing a phone call

To place a call on your phone, heed these steps:

1. **Open the Phone app.**

   The dialing operation on an Android phone is done by the Phone app or Dialer app. The app's launcher icon is found on the Home screen, often on the favorites tray. The app's icon features the Phone Handset icon, similar to the one shown in the margin.

2. **If necessary, display the dialpad.**

   Tap the Dialpad icon, shown in the margin, or tap a dialpad tab in the Phone app. A typical Dialpad screen is illustrated in Figure 5-1.

3. **Type a phone number.**

   You may hear the traditional touch-tone sounds as you punch in the number.

   If you make a mistake, tap the Delete icon, labeled in Figure 5-1, to back up and erase.

   The phone displays matching contacts as you type. Choose a contact to instantly input that person's number.

4. **Tap the Dial icon to place the call.**

   While a call is active, the screen changes to show contact information, or a contact image when one is available, similar to Figure 5-2.

5. **Place the phone to your ear and wait.**

6. **When the person answers the phone, talk.**

   What you say is up to you, though it's good not to just blurt out unexpected news: "I just ran over your cat" or "I think I just saw your daughter on the back of a motorcycle, leaving town at inappropriately high speeds."

   Use the phone's Volume key to adjust the volume during the call.

7. **Tap the End Call icon to end the call.**

   The phone disconnects. You hear a soft beep, which is the phone's signal that the call has ended.

You can do other things while you're making a call: Tap the Home navigation icon to run an app, read old email, check an appointment, or do whatever. Activities such as these don't disconnect you, although your cellular carrier may not allow you to do other things with the phone while you're on a call.

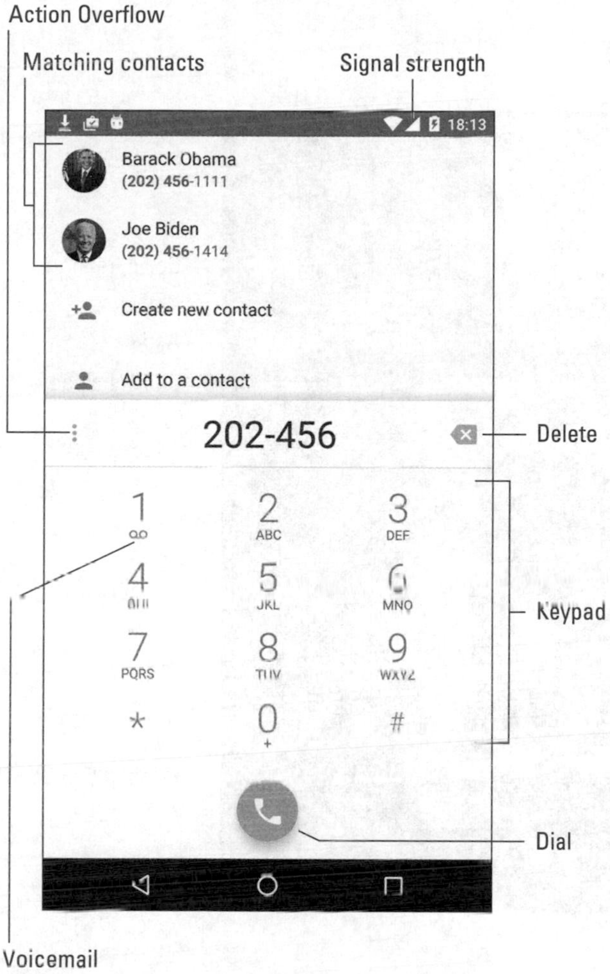

Action Overflow

Matching contacts

Signal strength

▼ ⊿ 🔋 18:13

Barack Obama
(202) 456-1111

Joe Biden
(202) 456-1414

+👤 Create new contact

👤 Add to a contact

⋮   202-456   ⊗ — Delete

1 ᴏᴏ	2 ABC	3 DEF
4 GHI	5 JKL	6 MNO
7 PQRS	8 TUV	9 WXYZ
*	0 +	#

Keypad

📞 — Dial

FIGURE 5-1:
The Phone
app's dialpad.

Voicemail

📞 To return to a call after doing something else, choose the Call in Progress notification icon, similar to the one shown in the margin.

>> Don't worry about the phone's microphone being too far away from your mouth; it picks up your voice just fine.

>> For hands-free operation, use earbuds with a microphone doodle. See Chapter 1 for details on obtaining and using this accessory.

>> You can connect or remove the earphones at any time during a call.

>> You can also use a Bluetooth headset to go hands-free. If the Bluetooth icon doesn't appear on the screen, tap the Speaker icon (refer to Figure 5-2) to ensure that the Bluetooth headset is active. See Chapter 17 for information on Bluetooth.

Mute

Speaker

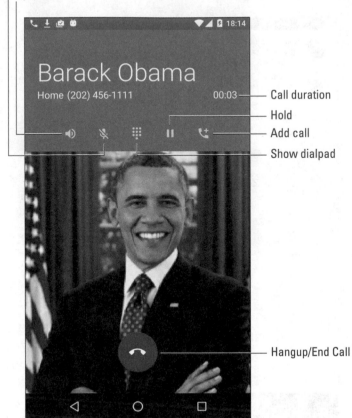

Barack Obama

Home (202) 456-1111   00:03 —— Call duration

—— Hold

—— Add call

—— Show dialpad

—— Hangup/End Call

FIGURE 5-2:
A successful
call.

>> If you're going hands-free, press the phone's Power/Lock key during the call to lock the phone. Locking the phone doesn't disconnect the call, but it does prevent you from accidentally touching the phone and hanging up or muting the call.

>> To mute a call, tap the Mute icon, shown earlier, in Figure 5-2. The Mute status icon, similar to the one shown in the margin, appears atop the touchscreen.

>> You can't accidentally mute or end a call when the phone is placed against your face; the phone's proximity sensor prevents that from happening.

>> Tap the Speaker icon and hold the phone at a distance to listen and talk, which allows you to let others listen and share in the conversation. The Speaker status icon appears when the speaker is active.

WARNING

» Don't hold the phone right at your ear while the speaker is active.

» If you're wading through one of those nasty voicemail systems, tap the Dialpad icon, labeled in Figure 5-2, so that you can "Press 1 for English" when necessary.

» See the later section "Multiple Call Mania," for information on using the Hold and Add Call icons.

» You hear an audio alert whenever the call is dropped or the other party hangs up. The disconnection can be confirmed by looking at the phone, which shows that the call has ended.

REMEMBER

» You cannot place a phone call when the phone has no service; check the signal strength. (Refer to Figure 5-1.) Also see the nearby sidebar, "Signal strength."

» You cannot place a phone call when the phone is in Airplane mode. See Chapter 22 for information.

» Also see Chapter 22 for details on International calling.

Dialing a contact

To access your phone's address book, start the Phone app and tap the Contacts tab, which might be titled All Contacts or feature an icon such as the one shown in the margin. Browse the list for someone to call; tap their entry and then tap their phone number or Phone icon to place the call.

SIGNAL STRENGTH

TECHNICAL STUFF

One of your Android phone's most important status icons Is Signal Strength. It appears in the upper right corner of the screen, next to the Battery status icon and the time.

The Signal Strength icon features the familiar bars, rising from left to right. The more bars, the better the signal. An extremely low signal is indicated by zero bars. When no signal is available, you may see a red circle with a line through it (the International No symbol) over the bars.

When the phone is out of its service area but still receiving a signal, you see the Roaming icon, which typically includes an *R* near or over the bars. See Chapter 22 for more information on roaming. Also see Chapter 17 for information on the mobile data network status icon.

>> You can also use the phone's address book app to find and phone a contact. See Chapter 7.

>> A special contact category in the phone's address book is Favorites. To quickly access your favorites, tap the Favorites tab in the Phone app. This tab may feature the Favorites (star) icon, shown in the margin.

>> Some variations on the Phone app place your favorite contacts on the main screen or on the Speed Dial screen. See the next section.

Using speed dial

To speed-dial a number, long-press one of the digits on the Phone app's dialpad. The phone number associated with that key is dialed instantly. For example, if your bookie is on speed dial 2, long-press the 2 key to instantly dial his number.

To assign a speed dial number, or just review the current settings, heed these steps when using the Phone app:

1. **Display the dialpad.**

2. **Tap the Action Overflow icon.**

   On some Samsung phones, tap the MORE button.

3. **Choose Speed Dial or Speed Dial Setup.**

   If you don't see that these are similar actions, the Phone app most likely lacks a speed dial feature.

   Most carriers configure number 1 as the voicemail system's number. The remaining numbers, 2 through 9, are available to program.

4. **Tap an item on the list, or tap the Add icon.**

   The item may say Add Speed Dial or Not Assigned as opposed to being blank.

5. **Choose a contact.**

6. **Repeat Steps 4 and 5 to assign more speed dial contacts.**

When you're done adding numbers, tap the Back navigation icon to exit the Speed Dial screen.

TIP

To remove a speed dial entry, long-press it and choose the Delete or Remove action. Or, in some cases, tap the Minus (Remove) icon to the right of the speed dial entry.

Adding pauses when dialing a number

When you tap the Phone icon to dial a number, the number is instantly spewed into the phone system, like water out of a hose. If you need to pause the number as it's dialed, you need to know how to insert secret pause characters. Two are available:

>> The comma (,) adds a 2-second pause.

>> The semicolon (;) adds a wait prompt.

To insert the pause or wait characters into a phone number, obey these directions:

1. **Type the number to dial.**

2. **At the point that the pause or wait character is needed, tap the Action Overflow icon.**

   On some phones, tap the MORE button.

3. **Choose the action Add 2-Sec Pause or Add Wait.**

4. **Continue composing the rest of the phone number.**

When the number is dialed and the comma (,) is encountered, the phone pauses two seconds and then dials the rest of the number.

When the semicolon (;) is encountered, the phone prompts you to continue. Tap the Yes or OK button to continue dialing the rest of the number.

TIP

>> The comma (,) and semicolon (;) can also be inserted into the phone numbers you assign to contacts in the phone's address book. See Chapter 7.

>> Alas, you cannot program an interactive phone number, such as one that pauses and lets you provide input and then continues to dial. You have to perform that task manually on an Android phone.

It's for You!

Who doesn't enjoy getting a phone call? It's an event! Never mind that it's the company that keeps calling you about lowering the interest rate on your credit cards. The point is that someone cares enough to call. Truly, your Android phone ringing can be good news, bad news, or mediocre news, but it always provides a little drama to spice up an otherwise mundane day.

Receiving a call

Several things can happen when you receive a phone call on your Android phone:

>> The phone rings or makes a noise signaling you to an incoming call.

>> The phone vibrates.

>> The touchscreen reveals information about the call, as shown at the top of Figure 5-3.

>> The car in front of you explodes in a loud fireball as your passenger screams something inappropriately funny.

That last item happens only in Bruce Willis movies. The other three possibilities, or a combination thereof, are your signals that you have an incoming call.

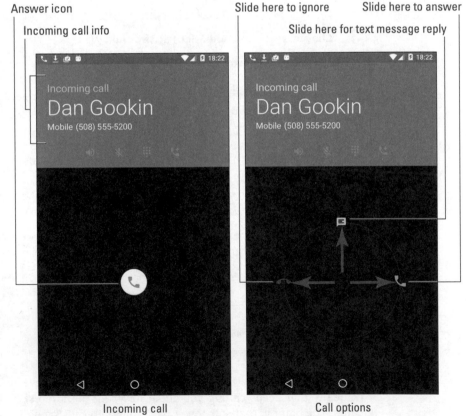

Answer icon

Incoming call info

Slide here to ignore

Slide here to answer

Slide here for text message reply

Incoming call

Call options

FIGURE 5-3:
You have an incoming call.

Work the touchscreen to answer the call. The stock Android method is shown in Figure 5-3: Swipe the Answer icon to the right to answer. Additional incoming call screens are illustrated in Figure 5-4. Generally, you swipe the Answer icon in a specific direction.

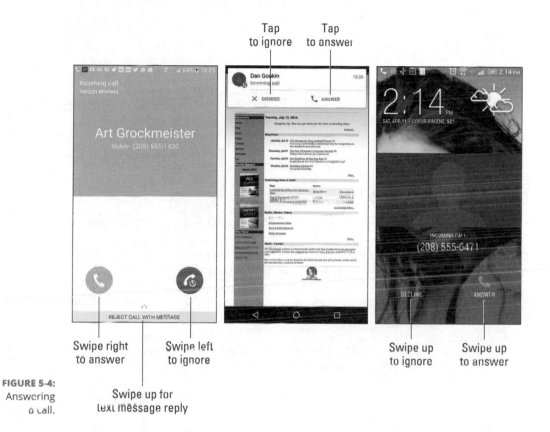

FIGURE 5-4:
Answering
a call.

The center image in Figure 5-4 shows how a call may look while you're using the phone: The incoming call appears as a card atop the screen. Tap the ANSWER button to take the call.

When you choose to answer the call, place the phone to your ear. If you're using earbuds or a Bluetooth headset, tap the button to answer. Say "Hello," or, if you're in a grouchy mood, say "What?" loudly.

Tap the End Call icon to hang up when you're done with the call. If the other party hangs up first, the call ends automatically.

>> You don't have to work a screen lock to answer a call; simply slide the Answer icon. If you want to do other things while you're on the call, you have to work the screen lock.

>> The contact's picture appears only when you've assigned a picture to the contact. Otherwise, a generic contact image shows up. (Refer to the far right side of Figure 5-4.)

>> The sound you hear for an incoming call is termed the *ringtone*. You can configure your phone's ringtone depending on who is calling, or you can set a universal ringtone. Ringtones are covered in Chapter 20.

Rejecting a call

Several options are available when you don't want to answer an incoming call.

Let the phone ring: Just do something else or pretend that you're dead. You can silence the ringer by pressing the phone's Volume key.

Dismiss the call: Swipe or tap the Decline icon, as illustrated earlier, in Figures 5-3 and 5-4.

Reply with a text message: Choose the text message rejection option, which sends the caller a text message but doesn't pick up the line.

In all cases, when you don't answer the phone, the call is sent to voicemail. See Chapter 6 for information on voicemail. Chapter 6 also covers the call log, which shows a list of recent calls incoming, missed, and rejected.

Rejecting a call with a text message

Rather than simply dismiss a call, you can reply to the call with a text message. Choose the Text Message icon or drawer when dismissing a call, shown in Figures 5-3 and Figure 5-4, respectively. Some phones may not display the text message rejection option until after you dismiss the call.

After choosing the text message rejection option, select a text message. Figure 5-5 shows how such a screen may look.

Incoming call info

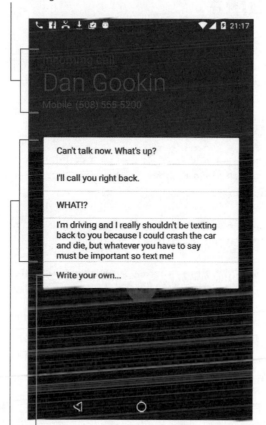

Can't talk now. What's up?

I'll call you right back.

WHAT!?

I'm driving and I really shouldn't be texting back to you because I could crash the car and die, but whatever you have to say must be important so text me!

Write your own...

FIGURE 5-5:
Text message rejection selection.

Create a new rejection message

Tap a rejection message

After you choose a text message, the incoming call is dismissed. In a few cellular seconds, the person who called receives the message.

>> Not every phone offers the text message rejection feature.

>> The method for adding, removing, or editing the call rejection messages differs from phone to phone. Generally, tap the Action Overflow icon or MORE button while using the Phone app. Choose Settings. Look for a Quick Responses or Call Rejection item.

>> See Chapter 8 for more information on text messaging.

Multiple Call Mania

As a human being, your brain limits your ability to hold more than one conversation at a time. Your phone's brain, however, lacks such a limitation. It's entirely possible for an Android phone to handle more than one call at a time.

Putting someone on hold

It's easy to place a call on hold — as long as your cellular provider hasn't disabled that feature. Tap the Hold icon, shown in the margin. Refer to Figure 5-2 for the Hold icon's location on the Call in Progress screen.

To take the call out of hold, tap the Hold icon again. The icon may change its look, for example, from a Pause symbol to a Play symbol.

Fret not if your phone's Call in Progress screen lacks the Hold icon. Rather than hold the call, mute it: Tap the Mute icon. That way, you can sneeze, scream at the wall, or flush the toilet and the other person will never know.

TIP

Receiving a new call when you're on the phone

You're on the phone, chatting it up. Suddenly, someone else calls you. What happens next?

Your phone alerts you to the new call, perhaps by vibrating or making a sound. Look at the touchscreen to see who's calling and determine what to do next. You have three options:

>> **Answer the call.** Slide the Answer icon just as you would answer any incoming call. The current call is placed on hold.

>> **Send the call directly to voicemail.** Dismiss the call as you would any incoming phone call.

>> **Do nothing.** The call eventually goes into voicemail.

When you choose to answer the second call, additional options become available to manage both calls. Use special icons on the Call in Progress screen to perform special, multi-call tricks:

Swap/Switch Calls: To switch between callers, tap the Swap or Switch Calls icon on the touchscreen. You might instead see a card at the bottom of the screen; tap the card to switch to that caller. The current person is placed on hold when you switch calls.

Merge Calls: To combine all calls so that everyone is talking (three people total), tap the Merge Calls icon. This icon may not be available if the merge feature is suppressed by your cellular provider.

End Call: To end a call, tap the End Call icon, just as you normally do. You're switched back to the other caller.

To end the final call, tap the End Call or Hang Up icon, just as you normally would.

> » The number of different calls your phone can handle depends on your carrier. For most subscribers in the United States, your phone can handle only two calls at a time. In that case, a third person who calls either hears a busy signal or is sent directly to voicemail.
>
> » If the person on hold hangs up, you may hear a sound or feel the phone vibrate when the call is dropped.
>
> » When you end a second call on the Verizon network, both calls may appear to have been disconnected. That's not the case: In a few moments, the call you didn't disconnect "rings" as though the person is calling you back. No one is calling you back, though; you're returning to that ongoing conversation.

Making a conference call

You can call two different people by using your Android phone's Merge Calls feature, also known as a conference call. Start by connecting with the first person, and then add a second call. Soon, everyone is talking. Here's how it works:

1. **Phone the first person.**

2. **After the call connects and you complete a few pleasantries, tap the Add Call icon.**

   The Add Call icon may appear as shown in the margin. If not, look for a generic Add (+) icon. After you tap that icon, the first person is placed on hold.

3. **Dial the second person.**

   You can use the dialpad or choose the second person from the phone's address book or the call history.

   Say your pleasantries and inform the party that the call is about to be merged.

4. **Tap the Merge or Merge Calls icon.**

   The two calls are now joined: Everyone you've dialed can talk to and hear everyone else.

5. **Tap the End Call icon to end the conference call.**

   All calls are disconnected.

REMEMBER

When several people are in a room and want to participate in a call, you can always put the phone in Speaker mode: Tap the Speaker icon on the ongoing call screen.

Your Android phone may feature the Manage icon while you're in a conference call. Tap this icon to list the various calls, to mute one, or to select a call to disconnect.

Forward Calls Elsewhere

One way to deal with incoming calls that doesn't require constant effort on your part is to forward the calls. The process is automatic: You choose what types of calls to forward and how to redirect them. You can forward all calls to another phone number or eternally banish individual contacts to voicemail.

Forwarding phone calls

The call forwarding feature lets you reroute incoming calls to your Android phone. For example, you can send all your calls to the office while you're on vacation. Then you have the luxury of making calls while ignoring anyone who calls you.

The Android operating system controls call forwarding options, though your cellular provider might control that feature instead. To determine which is which, follow these steps to configure call forwarding:

1. **Open the Phone app.**

2. **Tap the Action Overflow icon.**

3. **Choose Settings or Call Settings.**

4. **Choose Calls and then Call Forwarding.**

   On some phones, you can choose Call Forwarding without having to first choose the Calls item.

5. **Select a call forwarding option.**

   Sometimes only one option is available: a phone number to use for all incoming calls. You might instead see separate options, such as these:

   - *Always Forward:* All incoming calls are sent to the number you specify; your phone doesn't even ring. This option overrides all other forwarding options.

   - *When Busy:* Calls are forwarded when you're on the phone and choose not to answer. This option normally sends a missed call to voicemail.

   - *When Unanswered:* Missed calls are forwarded. Normally, the call is forwarded to voicemail.

   - *When Unreachable:* Calls are forwarded when the phone is turned off, out of range, or in Airplane mode. As with the two previous settings, this option normally forwards calls to voicemail.

6. **Set the forwarding number.**

   Or you can edit the number that already appears. For example, you can type your home number for the Forward When Unreached option so that your cell calls are redirected to your home number when you're out of range.

   To disable the option, tap the TURN OFF button.

7. **Tap the UPDATE button to confirm the new forwarding number.**

If your phone lacks call forwarding settings, you must rely upon the cellular carrier to set up and forward your calls. For example, Verizon in the United States uses the call forwarding options described in Table 5-1.

TABLE 5-1   ## Verizon Call Forwarding Commands

To Do This	Input First Number	Input Second Number
Forward unanswered incoming calls	*71	Forwarding number
Forward all incoming calls	*72	Forwarding number
Cancel call forwarding	*73	None

So, to forward all calls to (714) 555-4565, open the Phone app and dial the number ***727145554565**. You hear only a brief tone after dialing, and then the call ends. After that, any call coming into your phone rings at the other number.

REMEMBER

You must disable call forwarding to return to normal cell phone operations: Dial ***73**.

Call forwarding may affect your phone's voicemail service. See Chapter 6 for details.

Sending a contact directly to voicemail

You can configure your phone to forward any of your phone's address book contacts directly to voicemail. It's a great way to deal with a pest! Follow these steps:

1. Open the phone's address book app.

The app is called Contacts. See Chapter 7 for details.

2. Choose a contact, someone who's bothering you.

Display that person's contact information on the touchscreen.

3. Edit the contact: Tap the Edit icon or EDIT button.

On some phones, instead of editing the contact, tap the MORE button.

4. Tap the Action Overflow icon and choose All Calls to Voicemail.

The command might instead be titled Add to Reject List.

5. Save the changes: Tap the Done icon or SAVE button.

When a contact has been banished, you'll never receive another call from that person. Instead, you'll see a voicemail notification, but only when the caller chooses to leave you a message.

This feature is one reason you might want to retain contact information for someone with whom you never want to have contact.

Blocking calls

If your phone lacks an option to divert a contact's calls to voicemail, check for a Blocked Calls feature. Try these steps:

1. Open the Phone app.

2. Tap the Action Overflow icon or the MORE button.

3. Choose Settings.

4. Tap the Call Blocking item.

On some phones, you then choose Block List; otherwise, you see a list of numbers or contacts that are currently blocked.

5. Type a number to block, or choose an annoying number from the call log or contacts list.

You might also see a master control on the Block List screen that lets you banish any anonymous calls, which is a great way to deal with telemarketers.

Chapter 6

Missed Calls and Voicemail

The genius behind the mobile phone is that you can receive a call anywhere. I'm not implying that you'll never miss a call. In fact, some calls you'll miss on purpose: You whip out the phone, check the touchscreen to see who's calling, frown, and then dismiss the call, banishing it to voicemail. That happens a lot.

As a type of a computer, your Android phone dutifully keeps track of all calls, not only missed, but those you place and answer as well. Missed calls aren't sentenced to the digital ether, either: They can visit a place called voicemail. And that voicemail is more than just a "Leave your tone at the message" feature on an Android phone.

Who Called Who When?

All phone calls made on your Android phone are noted: To whom, from whom, date, time, duration, and whether the call was incoming, outgoing, missed, or dismissed. I call this feature the call log, although on your phone it may be referenced as Recent Calls or Call History. The place to look is in the Phone app.

Dealing with a missed call

 When you miss a call, the Missed Call notification icon appears atop the touch-screen. This icon, similar to the one in the margin, specifically applies to a call you didn't pick up; it doesn't appear for calls you dismiss or calls missed because the phone was turned off or in Airplane mode.

Choose the Missed Call notification to view the Phone app's call log, as illustrated in Figure 6-1. There you see details about who called and when. Choose an item from the call log to view more details and return the call.

Ignored call

Call History tab

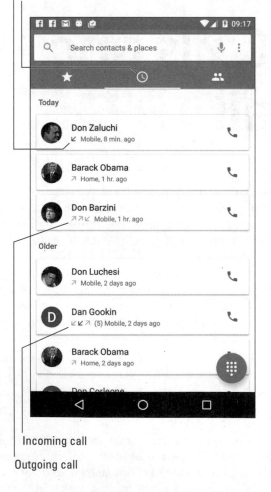

Incoming call

FIGURE 6-1:
The call log.

Outgoing call

See the next section for details on the call log.

REMEMBER

>> Some Android phones show more details on the Missed Call notification, including icons that let you return the call, text, and so on.

>> The phone doesn't consider a call you've dismissed as being missed.

Reviewing the call log

Your Android phone tracks all calls in a list I refer to as the call log. It's found in the Phone app: Tap the Call history tab, which might be titled Recent or Logs. Refer to Figure 6-1 to see what a call log might look like.

Swipe through the list to examine recent calls. The list is sorted so that the most recent calls appear at the top. Information is associated with each call, such as the date and time, the call duration, and whether the call was incoming, outgoing, missed, or ignored.

To see more details about the call, tap an item in the list. You can view more details, see a call history specific to that number or contact, or return the call.

TIP

>> The call log provides a quick way to create new contacts: Tap a phone number in the list, one that lacks contact information. Choose the action CREATE CONTACT or ADD TO CONTACTS. See Chapter 7 for more information about contacts.

>> To remove an item from the call log, get more information on the item: Tap it and choose INFO. On the next screen, tap the Delete (Trash) icon.

>> To clear the call log, tap the Action Overflow icon and choose the CLEAR CALL HISTORY or similar action. Some phones may require you to long-press an entry and then select all entries to clear the call log.

Boring carrier voicemail

The most basic, and most stupid, form of voicemail is the free voicemail service provided by your cell phone company. This standard feature has few frills and nothing that stands out differently for your nifty Android phone.

Carrier voicemail picks up missed calls as well as calls you thrust into voicemail. A notification icon, looking similar to the one shown in the margin, appears whenever someone leaves you a voicemail message. You can use the notification to dial into your carrier's voicemail system, listen to your calls, and use the phone's dialpad to delete messages or repeat messages or use other features you probably don't know about because no one ever pays attention to them.

Setting up carrier voicemail

If you haven't yet done it, set up voicemail on your phone. I recommend doing so even if you plan on using another voicemail service, such as Google Voice. That's because carrier voicemail, despite all my grousing, remains a valid and worthy fallback option when those other services fail.

Even if you believe your voicemail to be set up and configured, consider churning through these steps, just to be sure:

1. **Open the Phone app.**

2. **Tap the Action Overflow icon.**

    On some phones, tap the MORE button.

3. **Choose Settings.**

4. **Choose Calls or Call Settings.**

    Some phones may skip this step, showing a Voicemail item on the main Settings screen.

5. **Choose Voicemail or Voicemail Service.**

6. **Choose Service to determine which voicemail service the phone uses.**

    The options are either Carrier (or Your Carrier) and Google Voice, although only one option might be available.

When the Carrier option is chosen, the phone number used for voicemail is the carrier's voicemail service. You can confirm that number by choosing the Setup or Voicemail Settings item on the Voicemail screen.

TIP

After performing the steps in this section, call the carrier voicemail service to finish configuration: Long-press the 1 key in the Phone app's dialpad to access the voicemail system. Set your name, a voicemail password, a greeting, and various other steps as guided by the cellular provider's cheerful yet banal robot.

» Your phone may also feature a Voicemail app, which you can use to collect and review your messages.

REMEMBER

» The most important step for voicemail setup is to create a customized greeting. If you don't do so, you may not receive voicemail messages, or people may believe that they've dialed the wrong number.

Retrieving your messages

 When you have a voicemail message looming, a notification icon appears on the status bar, similar to the one shown in the margin. Choose this notification to connect to the voicemail service, or you can access the Phone app's dialpad where you long-press the 1 key.

What happens next depends on how your carrier has configured its voicemail service. Typically, you have to input your PIN. Afterward, the new messages play or you hear a menu of options. My advice: Look at the phone so that you can see the dialpad, and tap the Speaker icon so that you can hear the prompts.

To help you remember the prompts, write them down here:

Press _____ to listen to the first message.

Press _____ to delete the message.

Press _____ to skip a message.

Press _____ to hear the menu options.

Press _____ to hang up.

While you're at it, write your voicemail PIN: _____

The Wonders of Google Voice

Perhaps the best option I've found for working with voicemail is something called Google Voice. It's more than just a voicemail system: You can use Google Voice to make phone calls in the United States, place cheap international calls, and perform other amazing feats. In this section, I extol the virtues of using Google Voice as the voicemail system on your Android phone.

REMEMBER

>> Even when you choose to use Google Voice, I still recommend setting up and configuring the boring carrier voicemail, as covered earlier in this chapter.

>> With Google Voice configured as your phone's voicemail service, your voicemail messages arrive in the form of a Gmail message. The message's text is a transcription of the voicemail message.

>> Better than reading a Gmail message is using the Google Voice app to receive a Google Voice message. See the later section "Using the Google Voice app."

>> You may need to reset Google Voice after using call forwarding. That's because Google Voice relies upon call forwarding to deal with unanswered or dismissed calls. See Chapter 5 for more information on call forwarding.

Configuring Google Voice

To configure your Android phone for use with Google Voice, you must first create a Google Voice account. Start your adventure by visiting the Google Voice home page:

```
https://voice.google.com
```

I recommend using a computer to complete these steps: Follow the directions on the screen. Log in to your Google account, if necessary, and agree to the terms of service.

You're prompted to input information about the phone, and Google calls your cell phone to confirm the number and verify the service.

Eventually, you're given directions on how to forward missed calls to the Google Voice service. You see a number that you punch into the dialpad, which configures voicemail. I recommend that you write down that number for future reference or create a contact for the number. See Chapter 7 for information on creating new contacts.

Although this step configures a Google Voice account, you still need to program your phone to use Google Voice as its voicemail service. To best accomplish that task, install the Google Voice app. See the next section.

Using the Google Voice app

Google Voice transcribes your voicemail messages, turning the audio from the voicemail into a text message you can read. The messages all show up eventually in your Gmail inbox, just as though someone sent you an email rather than left you voicemail. It's a good way to deal with your messages, but not the best way.

The best way to handle Google Voice is to use the Google Voice app, available from Google Play. (See Chapter 16 for details on obtaining apps at Google Play.)

After the Google Voice app is installed, you have to work through the setup, which isn't difficult: The goal is to switch over the phone's voicemail number from the carrier's voicemail system to Google Voice. Eventually, you see the app's main interface, which looks and works similarly to an email program.

To review your voicemail messages, tap a message to read or play it, as illustrated in Figure 6-2.

Message text translation

Contact info (if available)

Incoming phone number

Turn on speaker

FIGURE 6-2:
Voicemail with the Google Voice app.

Play message

When new Google Voice messages come in, you see the Google Voice notification icon, as shown in the margin. To read or listen to the message, pull down the notifications and choose the item labeled Voicemail from *whomever*.

>> With Google Voice installed, you see two notices for every voicemail message: one from Google Voice and another for the Gmail message.

>> The Google Voice app works only after you acquire a Google Voice account and add your Android phone's number to that account. See the earlier section "Configuring Google Voice."

>> Google Voice tries its best at translation. As you can see from Figure 6-2, sometimes the translation isn't exact.

>> The text *Transcript Not Available* appears whenever Google Voice is unable to create a text message from your voicemail or whenever the Google Voice service is temporarily unavailable.

REMEMBER

Chapter 7

The Address Book

Once upon a time, humans went to the trouble of memorizing phone numbers. They didn't commit everyone's number to memory — just a few key contacts. So if a kid were stranded at the bowling alley, he could phone Mom's workplace, Grandpa, or even a neighbor lady to ask for a lift.

Today, it's your phone that memorizes phone numbers — and more. That's the duty of the phone's address book app, which also stores email addresses, physical addresses, and even random information such as birthdays and anniversaries. It's indispensable.

The People You Know

The address book app is central to many operations in an Android phone. It's used by Gmail, Email, Hangouts, and (most importantly) the Phone app. As a bonus, the address book app is probably full of people already; your Gmail contacts are instantly synchronized with the address book app, as are social networking contacts and any contacts associated with other accounts you've added to the phone.

Accessing the address book

The stock Android name for the phone's address book app is Contacts. It might also be called People on some phones. Look for the app's launcher icon on the Home screen or in the Apps drawer, as described in Chapter 3.

You can also access the list of contacts from within the Phone app: Tap the Contacts tab to view the list.

Figure 7-1 shows how the Contacts app might look. Its appearance on your phone may be different.

Swipe your finger on the touchscreen to scroll the list. You can use the index on the side of the screen (refer to Figure 7-1) to quickly scroll the list up and down. Large letters may appear as you scroll, to help you find your place.

Starred (favorite) contacts

All contacts

Add new contact

Index

FIGURE 7-1:
The Contacts
list.

When you locate a specific contact, tap the entry in the list. You see more information, similar to what's shown in Figure 7-2.

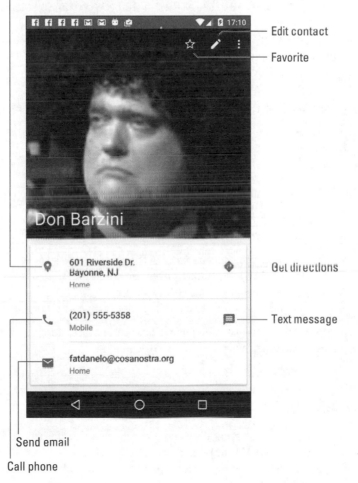

Locate in the Maps app

Edit contact

Favorite

Get directions

Text message

Don Barzini

601 Riverside Dr.
Bayonne, NJ
Home

(201) 555-5358
Mobile

fatdanelo@cosanostra.org
Home

FIGURE 7-2:
More details
about a
contact.

Send email

Call phone

The list of activities you can do with the contact depends on the information shown and the apps installed on the phone. Here are some options:

Place a phone call. To call the contact, tap one of the contact's phone entries. More than one might be shown, such as Home or Mobile.

Send a text message. Tap the Text Message icon (refer to Figure 7-2) to open the phone's text messaging app and send the contact a message. See Chapter 8 for information about text messaging.

Send email. Tap the contact's email address to compose an email message. Chapter 9 covers using email on your phone.

Locate the contact's address. When the contact's information shows a home or business address, tap that item to summon the Maps app and view the location. Refer to Chapter 12 to see all the fun stuff you can do with Maps.

Some tidbits of information that show up for a contact don't have an associated action. For example, the phone won't sing "Happy Birthday" when you tap a contact's birthday information.

>> Tap the Back icon to view the main address book list after viewing a contact's information.

>> Not every contact has a picture, and the picture can come from a number of sources (Gmail or Facebook, for example). See the later section "Adding a contact picture" for more information.

>> Many Android phones feature an account named Me. It shows your personal information as known by the phone. The Me account may be in addition to your other accounts shown in the address book.

>> Some cellular providers add accounts to your phone's address book. You may see entries such as BAL, MIN, or Warranty Center. Those aren't real people; they're shortcuts to various services. For example, the BAL contact is used on Verizon phones to get a text message detailing your current account balance.

>> Also see the later section "Joining identical contacts" for information on how to deal with duplicate entries for a single person.

Sorting the address book

The phone's address book displays contacts in a certain order, such as alphabetically by first name. If that presentation drives you nuts, change the order to whatever you like by following these steps in the address book app:

1. **Tap the Action Overflow icon, shown in the margin, and choose Settings.**

On some Samsung phones, tap the MORE button, and then choose Settings.

2. **Choose Sort By.**

3. **Choose First Name or Last Name, depending on how you want the Contacts list sorted.**

I prefer contacts sorted by first name, which is how the Contacts app normally does it.

4. **Choose Name Format.**

5. **Choose First Name First or Last Name First.**

   I prefer First Name First, which again is how the Contacts app presents the list.

The list of contacts is updated, displayed per your preferences.

Searching contacts

Rather than scroll your phone's address book with angst-riddled desperation, follow these steps:

1. **Tap the Search icon.**

   The Find Contacts text box appears.

   Some address book apps may show a Search text box atop the Contacts list without the need to tap the Search icon.

2. **Type a few letters of the contact's name.**

   As you type, the list of contacts narrows to match the search text.

3. **Tap a matching name to view the contact's information.**

To clear a search, tap the Close (X) icon at the right side of the search text box. To exit the search screen, tap the Back navigation icon.

Make New Friends

Having friends is great. Having more friends is better. You'll find myriad ways to add new friends and place new contacts into your phone. This section lists a few of the more popular and useful methods.

Creating a new contact from scratch

Sometimes it's necessary to create a contact when you meet another human being in the real world. In that case, you have more information to input, and it starts like this:

1. **Tap the Add New Contact icon in the Contacts app.**

   Refer to Figure 7-1 for the icon's look and location, although it may appear as a Plus Sign icon at the top of the screen.

2. **Choose an account with which to associate the contact.**

Account choices may be presented directly, or you may have to choose an account from an action bar. If you see your Google or Gmail address listed, you're good.

3. **Fill in the contact's information.**

Fill in the text fields with the information you know. The more information you provide, the better.

Some items feature an action bar from which you can choose options, such as whether a phone number is Work, Home, or Mobile.

To add a second phone number, an email address, or a location, tap the chevron or Plus Sign icon next to an item.

The Contacts app might sport a More Fields button. Use that button whenever you want to add more details for the contact, such as a birthday or website.

4. **Tap the Done icon or SAVE button to create the new contact.**

The new contact is created. As a bonus, it's also automatically synced with your Google account, or with whichever account you chose in Step 3.

TIP

>> Always type a phone number with the area code.

>> If you use an account other than Google/Gmail as your primary email account, choose it in Step 2. For example, if you use Yahoo!, choose that email account to store the new contact information.

WARNING

>> Do not choose the Device or Phone account in Step 2. When you do, the contact information is saved only on your Android phone. It isn't synchronized with the Internet or any other Android devices.

Adding a contact from the call log

A quick and easy way to build up your phone's address book is to add people as they phone you. To do so, check the call log:

1. **Open the Phone app.**

2. **Display the call log.**

Tap the History tab. See Chapter 6 for specific directions.

Unknown phone numbers appear by themselves, without a contact picture, name, or other details.

3. Display details about the phone number for an incoming call.

If the details aren't presented right away, long-press the entry or tap the Details button.

After displaying details about the unknown number you have two options: Add the number to an existing contact or create a new contact. Use either option to add a contact.

To add the unknown number to an existing contact, choose Update Existing. Locate the contact in the address book, and the phone number is automatically associated with that person or business.

Creating a new contact works just like creating a contact from scratch, with the exception that the phone number is already entered. Fill in other information: name, email, and what you know (though all that's really needed is the name). Save the new contact.

Creating a contact from an email message

You can use email messages to help build your phone's contacts list address book. That's because the email message supplies both the contact name and their email address. Follow these steps:

1. View the email message.

You can't add a contact from the inbox; tap the message to view its contents.

2. Tap the icon by the sender's name or tap the name itself.

The Gmail app uses a letter icon for unknown contacts, such as the H shown in the margin. The Email app may show a generic human icon, in which case you should tap that icon. Otherwise, tap the name itself.

3. Tap the Add User icon to create the new contact. Or, if prompted, choose the option to create a new contact.

The Add User icon is shown in the margin.

You might also see an option to update an existing contact. If so, choose that contact from the list to add the email address to their entry.

4. Add information about the new contact.

Refer to specific directions earlier in this chapter. The contact's email address is provided automatically.

5. Save or update the contact.

Tap the Done icon or SAVE button.

If you create a new contact, double-check the name. It might be added improperly, using the contact's email address or a name you don't normally associate with that contact.

Importing contacts from a computer

Your computer's email program is doubtless a useful repository of contacts you've built up over the years. You can export these valuable contacts into your Android phone. It's not the easiest thing to do, but it's possible.

The key is to save or export your computer email program's address book in the vCard (.vcf) file format. The vCard files can be read by the phone's address book app. The method for exporting contacts varies, depending on the email program:

> **For Microsoft Outlook,** you don't need to do a thing. Outlook contacts are automatically synchronized when you add your Exchange Server account to the phone. Refer to Chapter 2.

> **In the Windows Live Mail program,** choose Go ⇨ Contacts, and then choose File ⇨ Export ⇨ Business Card (.VCF) to export the contacts.

> **In Windows Mail,** choose File ⇨ Export ⇨ Windows Contacts, and then choose vCards (Folder of .VCF Files) from the Export Windows Contacts dialog box. Click the Export button.

> **On the Mac,** open the Contacts program and select the contacts you want to export. Choose File ⇨ Export ⇨ Export vCard to save the vCards as a single file.

After the vCard files are created, transfer them to your phone. Chapter 18 offers details on file transfer, which includes the direct cable connection as well as sharing the files on cloud storage.

I recommend that you copy the vCard files to the phone's Downloads folder.

With the vCard files stored on the phone, follow these steps in the Contacts app to complete the process:

1. Tap the Action Overflow icon or the MORE button.

2. Choose Import/Export.

> If you don't see this action, first choose Settings and then Import/Export Contacts.

3. **Choose Import from .vcf File.**

   This command may be titled differently, but the word *import* is in there somewhere.

4. **If prompted, allow the Contacts app to access the phone's files.**

5. **If prompted, direct the phone to use the Downloads folder as the source.**

   On some phones, you tap the IMPORT button and the vCard files are discovered automatically.

6. **Choose to save the contacts to your Google account.**

7. **If prompted, choose the option Import All vCard Files.**

   The imported contacts are synchronized to your Google account, or to whichever account you choose (Step 6), which instantly creates a backup copy.

The importing process may create some duplicates. That's okay: You can join two entries for the same person in your phone's address book. See the section "Joining identical contacts," later in this chapter.

Address Book Management

Sure, some folks just can't leave well enough alone. For example, some of your friends may change their phone numbers. They may move. They may finally get rid of their 20-year-old AOL email addresses. When such things occur, you must undertake the task of address book management.

Making basic changes

To make minor touch-ups on any contact, locate and display the contact's information in the address book app. Tap the Edit icon, which may look similar to the Pencil icon, shown in the margin. If this icon isn't available or obvious, tap the Action Overflow icon or MENU button and choose the Edit command.

Tap a field to edit information. Tap the Add icon or chevron next to an existing item to add another item, such as a second phone number or email address.

Some contact information cannot be edited. For example, fields pulled in from social networking sites can be edited only by that account holder on the social networking site.

When you're done editing, tap the Done icon or SAVE button.

Adding a contact picture

To spice up your phone's address book, consider assigning images to contacts. You can use an actual picture of the contact, a photo that reminds you of the contact, or something wholly inappropriate.

To use the phone's camera to set a contact image, follow these steps:

1. **Edit the contact's information.**

2. **Tap the contact's picture or the Change button that appears by the picture.**

   If nothing happens after tapping the picture, tap the Edit icon, and then tap the picture.

3. **Choose the Take Photo action.**

   The action might be titled Take Picture or something similar. On some phones, you may have to tap the X icon to first remove the current photo before you take a new one.

4. **Use the phone's camera to snap a picture.**

   Chapter 13 covers using the Camera app. Tap the Shutter icon to take the picture.

5. **Review the picture.**

   Not every phone lets you review the image — which is a good thing. If prompted, confirm that the image is up to snuff. If not, tap the Retry icon. Refer to Table 7-1 for definitions of the picture confirmation icons.

 6. **Tap the Done icon to confirm the new image and prepare for cropping.**

7. **Crop the image, as illustrated in Figure 7-3.**

   Adjust the cropping box so that it surrounds only the portion of the image you want to keep.

   In Figure 7-3, the image can also be rotated, which is a feature of the Photos app. See Chapter 13 for details.

8. **Tap the Done icon or SAVE button to crop and save the image.**

9. **Save your changes: Tap the Done icon or SAVE button.**

   With some Contacts apps, tap the Back navigation icon to exit Editing mode.

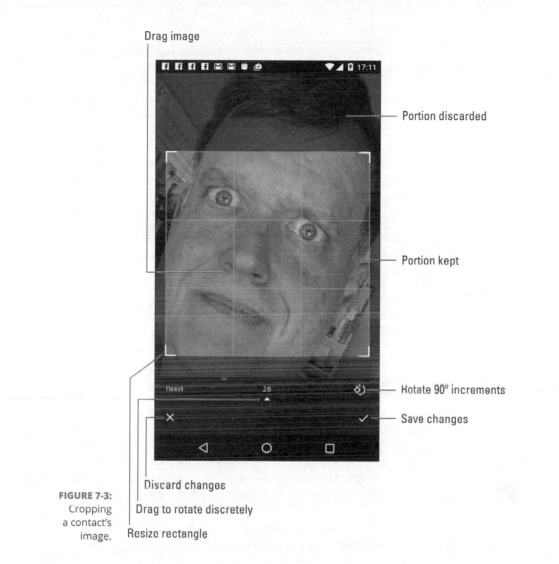

Drag image

Portion discarded

Portion kept

Rotate 90° increments

Save changes

Reset   28

Discard changes

Drag to rotate discretely

Resize rectangle

FIGURE 7-3:
Cropping
a contact's
image.

TABLE 7-1 Picture Confirmation Icons

Icon	Name	Function
↰	Retry	Take another photo
✓	Done	Accept the photo and crop
✗	Cancel	Abandon your efforts

The contact's image appears onscreen when the person calls, as well as when referenced in other apps, text messaging, Gmail, and so on.

To remove an image from a contact, follow Steps 1 and 2 in this section, but in Step 3 choose the action Remove Photo or tap the minus (–) icon.

When you choose the action Take Photo (refer to Step 3), you can choose a new image for the contact, such as one stored on the phone, accessed from your Google Drive, or obtained from any number of photo storage or management apps.

> >> A generic image or icon appears for contacts with an assigned photo or image.

> >> The contact image you see may be chosen by the contact themselves. Your Gmail contacts may show their own Google account image. Contacts associated with social networking sites may also use the person's own image instead of one you assign.

> >> Some images stored on the phone may not work for contact icons. For example, images synchronized with your online photo albums may be unavailable.

Making a favorite

A *favorite* contact is someone you stay in touch with most often. The person doesn't have to be someone you like — just someone you (perhaps unfortunately) phone often, such as your parole officer.

 To create a favorite, display a contact's information and tap the Star icon. When the star is highlighted, the contact is flagged as one of your favorites.

Favorite contacts appear on their own tab in the Contacts app, as well as in the Phone app. To quickly access a favorite, choose the person from the tab.

> >> To remove a favorite, tap the contact's star again. Removing a favorite doesn't delete the contact.

> >> A contact has no idea whether he's one of your favorites, so don't believe that you're hurting his feelings by not making him a favorite.

Joining identical contacts

Your phone can pull in contacts from sources such as Facebook, Gmail, Skype, and more. Because your friends also use multiple online services, you may discover several contact entries for the same person. When you notice such duplicates, you can *join* them. Here's how:

1. **Wildly scroll the address book until you locate a duplicate.**

   Well, maybe not wildly scroll, but locate a duplicated entry. Because the address book is sorted, duplicates may appear close together.

2. **Tap the Edit icon to edit one of the duplicate contacts.**

   Editing may not be required for some versions of the address book app.

   On some Samsung phones, tap the More icon and choose the Link Contacts action. Skip to Step 4.

3. **Tap the Action Overflow and choose Link or Join.**

   After choosing the action, you see a list of contacts that the phone guesses could be identical. It also shows all contacts, in case the guess is incorrect. Your job is to find the duplicate contact.

4. **Choose the duplicate contact from the list.**

   The contacts are merged, appearing as a single entry in the address book. On some phones you may need to tap a LINK button to complete the process.

Joined contacts aren't flagged as such in the address book, but you can easily identify them: A joined contact often lists two sources, such as Google and Facebook. Or you may see a Link icon (like a link in a chain).

TIP

Separating contacts

The topic of separating contacts has little to do with parenting, although separating bickering children is the first step to avoiding a fight. Contacts in the address book might not be bickering, but occasionally the phone may automatically join two contacts that aren't really the same person. When that happens, you can split them by following these steps:

1. **Display the improperly joined contact.**

   As an example, I'm Facebook friends with other humans named Dan Gookin. My phone accidentally joined my address book entry with another Dan Gookin.

2. **Tap the Action Overflow icon or MORE button.**

3. **Choose Separate.**

   The command might not be available, in which case you need to edit the contact, as described earlier in this chapter.

The Separate action might also be called Separate Contacts or Unlink Contacts. On some phones, the action is titled Manage Linked Contacts.

4. **Tap the OK button to confirm that you're splitting the contacts.**

   Or tap the minus (–) icon to remove one of the links. You may also have to save the contact.

You don't need to actively look for improperly joined contacts — you'll just stumble across them. When you do, feel free to separate them, especially if you detect any bickering.

Removing a contact

Every so often, consider reviewing your phone's contacts. Purge the folks whom you no longer recognize or you've forgotten. It's simple:

1. **Edit the forlorn contact.**

   In some cases you might find a Delete action on the Action Overflow menu or the MORE button's menu.

2. **Tap the Delete icon.**

   If you don't see the Delete icon (shown in the margin), tap the Action Overflow to look for a Delete action.

   If you can't find any other logical way to remove the content, it was brought in from another source, such as Facebook. You need to use that app to disassociate the contact.

3. **Tap the OK or DELETE button to confirm.**

   Poof! They're gone.

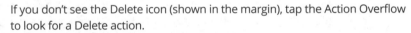

Because the phone's address book is synchronized with your Google account, the contact is also removed there.

WARNING

On the plus side, keep in mind that removing a contact doesn't kill the person in real life.

Chapter 8

Text Messaging Mania

Texting is the popular name for your Android phone's capability to send short, typed messages to another phone — specifically, another cell phone. The process echoes earlier technology, such as the telegraph and teletype. Interestingly enough, the acronym LOL, for laugh out loud, was first used in the 1880s:

```
Jeb: Stagecoach robbed again, but the boys
forgot the strongbox at the office. LOL.
```

Despite its seemingly anachronistic nature, texting remains a popular form of communications. Indeed, some young people text more than they use the phone to place a call. It's a great, convenient, and popular way to quickly communicate.

Msg 4U

Text messaging allows you to send short quips of text from one cell phone to another. As long as the other phone is on and receiving a signal, the message is received instantly. That makes texting a quick and worthy form of communication.

WARNING

>> Don't text while you're driving.

>> Don't text in a movie theater.

>> Don't text in any situation where it's distracting.

>> Most cell phone plans include unlimited texting; however, some older plans may charge you per text. Check with your cellular provider to be sure.

>> If you're over 25, you might want to know that the translation of this section's title is "Message for You."

TECHNICAL STUFF

>> The nerdy term for text messaging is SMS, which stands for Short Message Service.

Opening the texting app

Traditionally, Android phones come with one texting app. It's called Messenger, though other names include Messages, Messaging, and Text Messaging. You'll find this app's launcher on the Home screen, probably on the favorites tray. Tap the icon to open the text messaging app.

>> The Google Hangouts app can also handle text messages, though I recommend you stick with Messenger or the phone's default text messaging app. See Chapter 11 for more information on Google Hangouts.

>> Other apps may have messenger-like software, such as Facebook's Messenger program, which is used for Facebook chat and not text messaging.

Texting a contact

Here's the easy way to compose a text message:

1. **Open the phone's address book app.**

Refer to Chapter 7 for details on the address book app, usually named Contacts.

2. **Select a contact.**

WHETHER TO SEND A TEXT MESSAGE OR AN EMAIL

Sending a text message is similar to sending an email message. Both involve the instant electronic delivery of a message to someone else. And both methods of communication have their pros and cons.

The primary limitation of a text message is that it can be sent only to another cell phone. Email, on the other hand, is available to anyone who has an email address, whether or not the person uses a mobile communications device.

Text messages are pithy: short and to the point. They're informal, because the speed of reply is more important than trivia such as spelling and grammar. As with email, however, sending a text message doesn't guarantee a reply.

An email message can be longer than a text message. You can receive email on just about any Internet-connected device. Email message attachments (pictures, documents) are handled better and more consistently than text message (MMS) media.

Finally, email is considered a bit more formal than a text message. Still, when you truly desire formal communications, make a phone call or send a letter.

3. **Tap the Text Messaging icon, next to the phone number.**

The stock Android icon for text messaging is shown in the margin. The icon might also resemble an envelope.

Upon success, you see a text message window. Any previous conversation you've had appears on the screen, similar to what's shown in Figure 8-1.

4. **Type text in the Send Message or Enter Message text box.**

5. **Tap the Send icon to send the message.**

The Send icon may look like the one shown in the margin, or it might be the word *SEND*.

The message is delivered pretty much instantly, though getting an instant reply isn't guaranteed.

WARNING

>> You can send text messages only to cell phones. Aunt Ida cannot receive text messages on her landline that she's had since the 1960s.

>> Do not text and drive. Do not text and drive. Do not text and drive.

The person you're texting

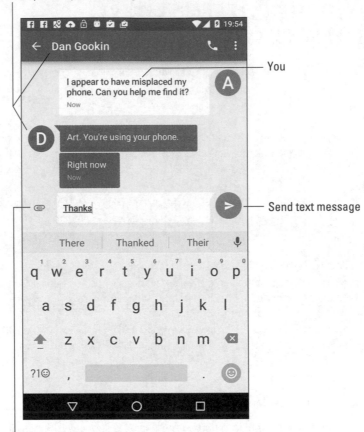

You

Send text message

FIGURE 8-1:
Sending a text
message.

Attachment

Composing a new text message

When you have only a phone number and you know it can receive text messages, follow these steps to send a text:

1. **Open the phone's texting app.**

2. **If it opens a specific conversation, tap the Back navigation icon to view the main screen.**

3. **Tap the Add icon to start a new conversation.**

The Add icon may look similar to what's shown in the margin, or it could be a Pencil icon or something similar.

4. **Type the phone number.**

   As you type, matching contacts appear. You can also type a contact name, if the person is already in the phone's address book.

5. **Type the message in the Send Message box.**

6. **Tap the Send icon or SEND button to send the message.**

Sending a text to multiple contacts

To send the message to multiple contacts, repeat the steps from the preceding section but in Step 4 continuing typing phone numbers or contact names. That's what makes the message a group text.

WARNING

When you receive a group text message (one that has several recipients), you can choose whether to reply to everyone. Look for the Reply All icon or button when composing your response. But please use caution when replying to everyone. Many people don't like group text messages, because they can go on forever and be horribly annoying.

Continuing a text message conversation

The text messaging app keeps track of old conversations, and you can pick up where you left off at any time: Open the texting app, peruse the list of existing conversations, and tap one to review what has been said or to begin something new.

Receiving a text message

New text messages are heralded by a notification atop the screen, similar to the one shown in the margin. If the phone is on, you may even see a card slide in with the message, as illustrated in Figure 8-2.

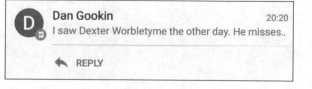

FIGURE 8-2: A new message has arrived.

COMMON TEXT MESSAGE ABBREVIATIONS

Texting isn't about proper English. Indeed, many of the abbreviations and shortcuts used in texting, such as LOL and BRB, are slowly becoming part of the English language.

The weird news is that these acronyms weren't invented by teenagers. Sure, the kids use them, but the acronyms have their roots in the Internet chat rooms of yesteryear. Regardless of its source, you might find using a shortcut handy for typing messages quickly. Or maybe you can use this reference for deciphering an acronym's meaning. You can type acronyms in either uppercase or lowercase letters.

2	To, also		OMG	Oh my goodness!
411	Information		PIR	People in room (watching)
BRB	Be right back		POS	Person over shoulder (watching)
BTW	By the way			
CYA	See you		QT	Cutie
FWIW	For what it's worth		ROFL	Rolling on the floor, laughing
FYI	For your information		SOS	Someone over shoulder (watching)
GB	Goodbye			
GJ	Good job		SMH	Shaking my head
GR8	Great		TC	Take care
GTG	Got to go		THX	Thanks
HOAS	Hold on a second		TIA	Thanks in advance
IC	I see		TMI	Too much information
IDK	I don't know		TTFN	Ta-ta for now (goodbye)
IMO	In my opinion		TTYL	Talk to you later
JK	Just kidding		TY	Thank you
K	Okay		U2	You too
L8R	Later		UR	You're, you are
LMAO	Laughing my ass off		VM	Voicemail
LMK	Let me know		W8	Wait
LOL	Laugh out loud		XOXO	Hugs and kisses
NC	No comment		Y	Why?
NP	No problem		YW	You're welcome
NRN	No reply needed (necessary)		ZZZ	Sleeping

If you see the message card, tap it to view the message and tap the REPLY button. Some phones may show more buttons, which offer more control. Otherwise, you can ignore the message card and it goes away. The message itself, and the notification icon, remain until you view the message.

>> Choose the new text message notification to view any new text message.

>> The Messenger (or similar) app may show a counter flag, similar to what's shown in the margin, indicating unread text messages.

Forwarding a text message

It's possible to forward a text message, but it's not the same as forwarding email: Your phone lets you forward only the information in a text messaging cartoon bubble, not the entire conversation. Here's how it works:

1. If necessary, open a conversation in the phone's texting app.

2. Long-press the text entry (the cartoon bubble) you want to forward.

3. Choose Forward or Forward Message.

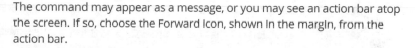

The command may appear as a message, or you may see an action bar atop the screen. If so, choose the Forward Icon, shown in the margin, from the action bar.

From this point on, forwarding the message works like sending a new message from scratch: Choose a recipient or type their contact name or phone number. The text from the cartoon bubble you selected (refer to Step 2) is pasted into the Send Message text box. Tap the Send icon or SEND button to forward the message.

Other icons that may appear on the action bar (refer to Step 3) include Share, Copy, Info, and Delete (trash).

Multimedia Messages

The term *texting* sticks around, yet a text message can contain media — usually a photo — although short videos and audio can also be shared with a text message. Such a message ceases to be a mere text message and becomes a multimedia message.

» Multimedia messages are handled by the same app you use for text messaging.

» Not every mobile phone can receive multimedia messages. Rather than receive the media item, the recipient may be directed to a web page where the item can be viewed on the Internet.

» The official name for a multimedia text message is Multimedia Messaging Service, abbreviated MMS.

Creating a multimedia text message

As with other things on your Android phone, you need to think of sharing when it comes to attaching media to a text message. Obey these steps:

1. **Open the app that contains or shows the item you want to share.**

   For example, open the Photos app to view a picture or view a page in the web browser app.

2. **View the item and tap the Share icon.**

3. **Choose the phone's text messaging app from the list of apps.**

4. **Continue sending the text message as described earlier in this chapter.**

It's also possible to attach media to a message from within the text messaging app. To do so, begin composing the message, but look for the Attachment icon, similar to the one shown in the margin. (Also refer to Figure 8-1.) Tap that icon, and then choose the media to attach.

In just a few, short, cellular moments, the receiving party will enjoy your multimedia text message.

Receiving a multimedia message

A multimedia attachment comes into your phone just like any other text message. You may see a thumbnail preview of whichever media was sent, such as an image, a still from a video, or the Play icon to listen to audio. To preview the attachment, tap it.

To do more with the multimedia attachment, long-press it and then select an action from the list. For example, to save an image attachment, long-press the image thumbnail and choose Save Picture.

Text Message Management

You don't have to manage your messages. I certainly don't. But the potential exists: If you ever want to destroy evidence of a conversation, or even do something as mild as change the text messaging ringtone, it's possible.

Removing messages

Although I'm a stickler for deleting email after I read it, I don't bother deleting my text message threads. That's probably because I have no pending divorce litigation. Well, even then, I have nothing to hide in my text messaging conversations. If I did, I would follow these steps to delete a conversation:

1. **Open the conversation you want to remove.**

    Choose the conversation from the main screen in your phone's text messaging app.

2. **Tap the Action Overflow and choose Delete.**

3. **Tap the DELETE button to confirm.**

    The entire conversation is gone.

If these steps don't work, an alternative is to open the main screen in the text messaging app and long-press the conversation you want to zap. Tap the Delete button, and then tap the DELETE or OK button to confirm.

Individual cartoon bubbles can be removed from a conversation: Long-press the bubble and then tap the Trash icon or the DELETE button.

Setting the text message ringtone

The sound you hear when a new text message floats in is the text message ringtone. It might be the same sound you hear for all notifications, though on some Android phones it can be changed to something unique.

Follow these steps in the Messenger app to set a new text message ringtone:

1. **Tap the Action Overflow.**

2. **Choose People & Options.**

3. **Choose Sound.**

4. **Select a sound from the list and tap OK.**

On some Samsung phones, open the Messages app and follow these steps:

1. **At the main screen, tap the MORE button.**

2. **Choose Settings.**

3. **Choose Notifications.**

4. **On the Notifications screen, choose Notification Sound.**

5. **Choose a sound from the list.**

6. **Touch the Back navigation button when you're pleased with the new sound.**

On other phones, the new message ringtone might be set from the Settings app. Choose Sound & Notification and look for an item specific to the text message ringtone.

Chapter 9
Email This-and-That

The first official telegraph message was, "What hath God wrought?" The first telephone call was, supposedly, "Mr. Watson. Come here. I want you." The first email message reportedly said, "Wànt tø büy somé chéàp Ciàlis?"

You'll be delighted to know that your phone can send and receive email. You can ask questions about the Almighty, receive requests for assistance, and trade in pharmaceuticals.

Email on Your Android Phone

For most Android phones, two apps handle email: Gmail for Google email, and the Email app for everything else — including web-based email such as Yahoo! Mail or Windows Live mail, or corporate email.

Recently, Google has expanded the Gmail app to handle all types of email. This change may be evident on your phone, or you might still find separate Gmail and

Email apps. The good news is that both apps work in a similar manner, so your sanity should remain more or less intact.

>> The Gmail app is updated frequently. To review any changes since this book went to press, visit my website at

```
www.wambooli.com/help/android
```

>> Although you can use the phone's web browser app to visit the Gmail website, you should instead use the Gmail app to pick up your Gmail. Likewise, you can access a web page to read most other email, but instead use the Email app.

>> If you forget your Gmail password, visit this web address:

```
www.google.com/accounts/ForgotPasswd
```

Setting up the first email account

The first email account on your phone is your Gmail account, which is required in order to use an Android phone. After that, you add email accounts as described in the next section.

If your phone uses the Email app, setting up the first email account works differently from adding additional accounts. You must know the email account's sign-in (your email address) and password. When you have that information, follow these steps in the Email app to configure that first account:

1. Start the Email app.

Look for it in the Apps drawer. If it's not there, use the Gmail app instead. See the next section.

The first screen you see is Set Up Email. If you've already run the Email app, you're taken to the Email inbox and you can skip these steps. See the next section for information on adding accounts.

2. Type the email address you use for the account.

For example, if you have a Comcast email account, use the onscreen keyboard to type your whoever@comcast.net email address in the Email Address box.

If you see the .com (dot-com) key on the onscreen keyboard, use it to more efficiently type your email address.

3. Type the account's password.

4. **Tap the NEXT button or, if you can't see that button, tap the DONE button on the onscreen keyboard.**

   If you're lucky, everything is connected and you can move on to Step 5. Otherwise, you have to specify the details as provided by your ISP. See the later section "Adding an account manually."

5. **Set account options.**

   You might want to reset the Inbox Checking Frequency option to something other than 15 minutes. I recommend keeping the other items selected until you become familiar with how your phone handles email.

6. **Tap the NEXT button.**

7. **Give the account a name and confirm your own name.**

   The phone chooses the email server name or your email address as the account name. If that choice doesn't ring a bell, rename the account. For example, I name my ISP's email account Main because it's my main account.

   The Your Name field shows your name as it's applied to outgoing messages. So if your name is really Annabelle Leigh Meriwether and not amer82, you can make that change now.

8. **Tap the DONE button.**

   You're done.

The next thing you see is your email account inbox. The phone proceeds to synchronize any pending email you have in your account, updating the screen as you watch. See the later section "You've Got Mail" for what to do next.

TIP

If you use Yahoo! Mail, I recommend getting the Yahoo! Mail app, which handles your Yahoo! mail far better than the Email app (or the Gmail app). The Yahoo! Mail app gives you access to other Yahoo! features that you may use and enjoy. Obtain the Yahoo! Mail app from Google Play. See Chapter 16.

Adding more email accounts

The Email app, as well as the newest version of Gmail, can be configured to pick up email from multiple sources. If you have a Windows Live account or maybe an Evil Corporate Account in addition to your ISP's account, you can add them to the phone's email account inventory.

Both the Email and Gmail apps offer different ways to add a new email account. A better approach is to use the Settings app. Obey these directions:

1. **Open the Settings app.**

   It's found in the Apps drawer; tap the Apps icon on the Home screen to view the Apps drawer.

2. **Choose Accounts.**

   On some Samsung phones, tap the General tab to locate the Accounts item.

3. **Tap Add Account.**

4. **Choose the email account type.**

   The three options for adding email accounts are described in this list:

   Exchange or Microsoft Exchange ActiveSync: For a corporate email account hosted by an Exchange Server (Outlook mail)

   Personal (IMAP): For web-based email accounts, such as Microsoft Live

   Personal (POP3): For traditional, ISP-email accounts, such as Comcast

   See the later section "Adding a corporate email account" for information on the Exchange option. The other two options work pretty much the same. In fact, some phones may have a single Email option instead of the IMAP and POP variations.

5. **Type your email address and tap the NEXT button.**

6. **Type the email account password and tap the NEXT button.**

7. **Continue working through the email setup as prompted on the screen.**

   Refer to the preceding section, after Step 5 (in that section).

The big difference between creating the first email account and adding more is that you're asked whether the account you just added is the primary or default account. See the later section "Setting the primary email account" for details.

The new email account is synchronized immediately after it's added, and you see the inbox. See the later section "Checking the inbox."

Adding an account manually

TECHNICAL
STUFF

If your email account isn't recognized, or some other boo-boo happens, you have to manually add the account. The steps in the preceding sections remain the same, but you need to provide more specific and technical particulars. This information includes tidbits such as the server name, port address, domain name, and other bothersome details.

My advice is to contact your ISP or email provider: Look on their website for specific directions on adding their email service to an Android phone. Phone them if you cannot locate specific information on the website.

The good news is that manual setup is very rare these days. Most ISPs and webmail accounts are added painlessly, as described earlier in this chapter.

Adding a corporate email account

TIP

The easiest way to set up your evil corporation's email on your phone is to have the IT people do it for you. Or you may be fortunate, and on the organization's intranet you'll find directions. I present this tip because configuring corporate email, also known as Exchange Server mail, can be a difficult and terrifying ordeal.

It's possible to add the account on your own, but you still need detailed information. Specifically, you need to know the domain name, which may not be the same as the outfit's website domain. Other details may be required as well, such as port numbers or authentication such-and-such.

Above all, you need to apply a secure screen lock to your phone to access Outlook email. This means you need to add a PIN or password to the device, which is covered in Chapter 21. You cannot access the Exchange Server without that added level of security.

And there's more!

You also need to grant Remote Security Administration privileges. This means your organization's IT gurus will have the power to remotely wipe all information from your phone. You must activate that feature, which is part of the setup process.

The bonus is that when you're done, you have full access to the Exchange Server info. That includes your email messages as well as the corporate address book and calendar. Refer to Chapter 7 for information on the address book app; the Calendar app is covered in Chapter 15.

You've Got Mail

New email arrives into your phone automatically, picked up according to the Gmail and Email apps' synchronization schedules. If Gmail is the phone's only email app, you use it to read all your email. Otherwise, use the Email app to read non-Gmail email.

Getting a new message

You're alerted to the arrival of a new email message in your phone by a notification icon. The icon differs depending on the email's source.

 For a new Gmail message, the New Gmail notification appears, similar to the one shown in the margin.

 For a new email message, you see the New Email notification.

For email you receive from an Exchange Server, or corporate email, you might even see the New Exchange Mail notification.

Conjure the notifications drawer to review the email notifications. Tap a notification to be whisked to an inbox for instant reading.

Checking the inbox

To peruse your Gmail, start the Gmail app. A typical Gmail inbox is shown in Figure 9-1.

Choose non-Gmail accounts by tapping the account icon (the bubble) from the top of the navigation drawer, illustrated in Figure 9-1. You can view only one account's inbox at a time.

If your phone uses the Email app, open it to view its inbox. Either you see a single account's inbox or you can choose to view the universal inbox, shown as Combined Inbox in Figure 9-2.

REMEMBER

>> Gmail doesn't show up in the Email app, not even in the Combined inbox. Use the Gmail app to read your Google mail.

>> The Gmail app lacks a combined inbox. To view specific inboxes, tap an account bubble on the navigation drawer. (Refer to Figure 9-1.)

>> Multiple email accounts that are gathered in the Email app are color coded. When you view the combined inbox, you see the color codes to the left of each message, as shown in Figure 9-2.

>> To view an individual account's inbox in the Email app, choose the account from the action bar or from the navigation drawer.

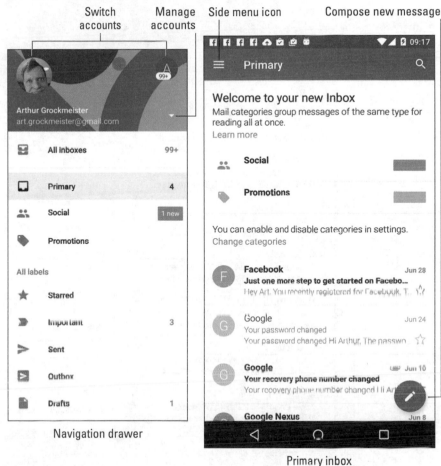

Switch accounts   Manage accounts   Side menu icon   Compose new message

Navigation drawer

Primary inbox

FIGURE 9-1:
The Gmail
inbox.

Reading email

To view a specific email message, tap its entry on the inbox screen, as shown earlier, in Figures 9-1 and 9-2. Or you can choose a new email notification. Reading and working with the message operates much the same whether you're using the Gmail app or Email app:

 Use your finger to swipe the message up or down.

>> Browse between messages by swiping the screen left or right.

>> Tap the left-pointing arrow or chevron at the top left corner of the screen to return to the inbox.

Attachment

Combined inbox ▼   SEARCH   MORE ——— Action bar

Unread 50

✉ **WIDMYER, STEVE**   09:18
Dallas . . . .
In the wake of the national tragedies that... ⚑

✉ **Elderly Tubs**   08:44
SPAM: Walk In Bathtubs - Don't be afraid...
Enjoy an easier and safer bathing experie... ☆

✉ HAMMOND, JIM   08:27
🖇 absence
I will be out of the office July 14 to July 19. ⚑

✉ **CafePress**   06:25
Summer is here! Play outside and save 2...
@font-face { font-family: 'Open Sans'; font... ☆

✉ **Foscam.US**   04:11
Sale on HD 1080p WiFi Outdoor IP Securi...
Securing Your Foscam Cameras As a Fos... ☆

✉ **Filtered**   02:50
Workforce productivity webinar with Don...
Register for free today | WORKFORCE PR... ☆

✉ **MacMall**   00:05
Apple TV 64GB $179 | HP 19" Monitor $6 ✎
To ensure you receive our emails, please... ——— Compose new message

FIGURE 9-2:
Messages in
the Email app.   Color-coded account categories

To work with the message, use the icons that appear above or below the message text. These icons, which may not look exactly like those shown in the margin, cover common email actions:

 Reply: Tap this icon to reply to a message. A new message window appears with the To and Subject fields reflecting the original sender(s) and subject.

 Reply All: Tap this icon to respond to everyone who received the original message, including folks on the Cc line. Use this option only when everyone else must receive a copy of your reply.

 Forward: Tap this icon to send a copy of the message to someone else.

Delete: Tap this icon to delete a message. This icon is found atop the message.

If you don't see all these icons — specifically, the Reply All and Forward icons — change the phone's orientation to horizontal. You can also tap the Action Overflow icon to locate the email actions.

TIP

If you've properly configured the Email program, you don't need to delete messages you've read. See the section "Configuring the server delete option," later in this chapter.

Make Your Own Email

Although I use my phone often to check email, I don't often use it to compose messages. That's because most email messages don't demand an immediate reply. When they do, or when the mood hits me and I feel the desperation that only an immediate email message can quench, I compose a new email message on my Android phone using the methods presented in this section.

Writing a new electronic message

Creating a new email epistle works similarly with both the Gmail and Email apps. The key is to tap the Compose icon, illustrated in Figures 9-1 and 9-2 for the Gmail and Email apps, respectively. Tap that icon and then fill in the blanks, adding recipient, subject, and message text.

Figure 9-3 shows the layout of the typical email composition screen, which should be familiar to anyone who's used email on a computer. The screen is similar between the Gmail and Email apps, although not identical.

To send the message, tap the Send icon, illustrated in Figure 9-3 and shown in the margin.

REMEMBER

>> To choose an email account for sending, tap the chevron next to your name atop the Compose screen, shown in Figure 9-3. Otherwise, the primary or default email account is used. See the later section "Setting the primary email account."

>> You need only type a few letters of the recipient's name. Matching contacts from the phone's address book appear in a list. Tap a contact to automatically fill in the To field.

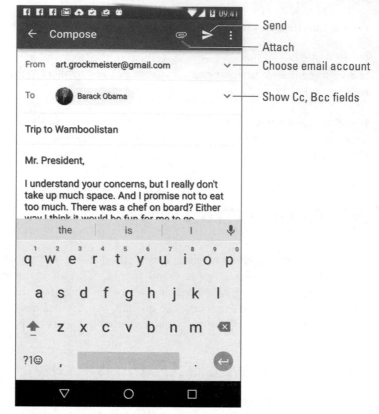

FIGURE 9-3:
Writing a new
email message.

>> To view the Cc and Bcc fields, tap the chevron, shown in Figure 9-3. If that doesn't work, tap the Action Overflow and choose Add Cc/Bcc.

>> To cancel a message, tap the Action Overflow and choose Discard. Tap the OK or Discard button to confirm.

>> To save a message, tap the Action Overflow and choose the Save Draft action. Drafts are saved in the Drafts folder. You can open unfinished email in that folder for further editing or sending.

>> Some phones feature a formatting toolbar in the Email app. Use it to apply basic text formatting to the message's text.

Sending email to a contact

A quick and easy way to compose a new message is to find a contact in the phone's address book app. Heed these steps:

1. **Open the phone's address book app.**

   See Chapter 7 for details on using this app.

2. **Locate the contact to whom you want to send an electronic message.**

3. **Tap the contact's email address.**

4. **Choose Gmail to compose the message.**

   If the Email app is available, or another app such as Yahoo! Mail, choose it instead. Also see Chapter 19 for information on choosing a default app.

At this point, creating the message works as described in the preceding sections.

Message Attachments

 The key to understanding email attachments on your phone is to look for the paperclip icon. When you see that icon, you can either deal with an attachment for incoming email or add an attachment to outgoing email.

Dealing with an attachment

Attachments are presented differently between the Gmail and Email apps. Either way, your goal is the same: to view the attachment or to save it. Sometimes you can do both!

Figure 9-4 shows how the Gmail and Email apps can present an email attachment. To deal with the attachment, tap it. In most cases, the attachment opens an appropriate app. For example, a PDF attachment might open in the Quickoffice app.

Potential actions you can perform with an attachment include

Preview: Open the attachment for viewing.

 Save to Google Drive: Send a copy of the attachment to your Google Drive (cloud) storage.

Save: Save the attachment to the phone's storage.

Download Again: Fetch the attachment from the mail server.

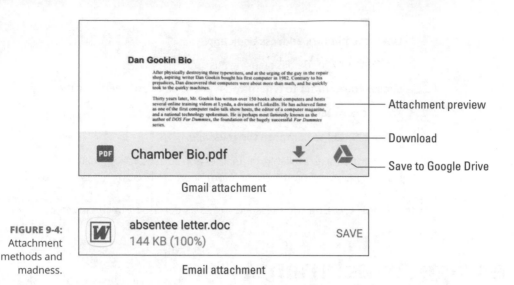

Dan Gookin Bio

After physically destroying three typewriters, and at the urging of the guy in the repair shop, aspiring writer Dan Gookin bought his first computer in 1982. Contrary to his prejudices, Dan discovered that computers were about more than math, and he quickly took to the quirky machines.

Thirty years later, Mr. Gookin has written over 150 books about computers and hosts several online training videos at Lynda, a division of LinkedIn. He has achieved fame as one of the first computer radio talk show hosts, the editor of a computer magazine, and a national technology spokesman. He is perhaps most famously known as the author of *DOS For Dummies*, the foundation of the hugely successful *For Dummies* series.

———— Attachment preview

PDF Chamber Bio.pdf ⬇ △ ———— Download
———— Save to Google Drive

Gmail attachment

FIGURE 9-4:
Attachment
methods and
madness.

W absentee letter.doc
144 KB (100%)   SAVE

Email attachment

As with email attachments received on a computer, the only problem you may have is that the phone lacks the app required to deal with the attachment. When an app can't be found, you either have to suffer through not viewing the attachment or reply to the message and direct the sender to resend the attachment in a common file format.

>> Common file formats include PNG and JPEG for pictures, and HTML or RTF for documents. PDF and DOCX for Adobe Acrobat and Microsoft Word documents are also common.

>> See Chapter 19 for information on selecting a default app, should the phone prompt you to choose an app to view the attachment.

>> Images in an email are technically attachments, although they're sometimes not displayed. Tap the message's SHOW PICTURES button to see the images.

TECHNICAL
STUFF

>> Attachments are saved in the Downloads or Download folder on the phone's internal storage. Use the Downloads app to view that folder's contents, as covered in Chapter 10. Also see Chapter 18 for information on phone storage.

Sharing an attachment

You can attach an item to any message you create, but on an Android phone the preferred way is to start with the source media or item and then use the Share icon. Follow these steps:

1. **Open the app that created the attachment, or in which you can view the attachment you want to send.**

   For example, to send a photo, open the Photos app. To send a YouTube video link, open the YouTube app.

2. **View the item you want to share.**

3. **Tap the Share icon.**

   A list of apps appears.

4. **Choose Email or Gmail.**

   This prompt may not appear if you've chosen a default email app.

5. **Compose the message.**

   The item is attached to the message, so at this point, writing the email works just as described earlier in this chapter.

 The computer way of attaching a file to a message is to first use the Gmail or Email apps and write the message. Tap the Action Overflow icon and choose the command Attach File, or tap the Attachment icon, shown in the margin. Use the screen that appears to choose an app, or browse the files stored on the phone to pick one as an attachment.

» The variety of items you can attach depends on which apps are Installed on the phone.

» The Gmail and Email apps sometimes accept different types of attachments. If you cannot attach something by using the Gmail app, try using the Email app instead.

Email Configuration

You can have oodles of fun and waste oceans of time confirming and customizing the email experience on your Android phone. The more interesting things you can do are to modify or create an email signature, specify whether messages retrieved on the phone are deleted from the server, and assign a default email account.

Creating a signature

I highly recommend that you create a custom email signature for sending messages from your phone. Here's my signature:

```
DAN
This was sent from my Android phone.
Typos, no matter how hilarious, are unintentional.
```

To create a custom signature for your email accounts, obey these directions:

1. **Open the Settings app**

2. **Choose Accounts.**

    Some Samsung phones hide the Accounts item on the General tab.

3. **Choose an email account from the list.**

    For example, choose the Personal (IMAP) or Personal (POP3) items, or tap the Email item if you see only it.

4. **Choose Account Settings.**

    This item might be titled Settings.

5. **Choose a specific email account.**

6. **Choose Signature.**

    Any existing signature appears on the card, ready for you to edit or replace it.

7. **Type or dictate your signature, and then tap the OK button.**

    The signature you set is appended automatically to all outgoing email you send from that account.

You need to repeat these steps (actually, only Steps 5 through 7) for each of your email accounts.

TIP

If you want to use the same signature on other accounts, select and copy the text from Step 7. See Chapter 4 for details on text editing.

These steps may not work on your phone's Email app. If so, open that app and locate the Settings item. (Tap the Action Overflow or the MORE button.) Choose your email account from the list, and then choose Signature to create a custom signature.

Configuring the server delete option

Unlike reading your email on a computer, messages you fetch from traditional ISPs (such as Comcast or CenturyLink) on your phone aren't deleted from the email server. The advantage is that you can later use a computer to retrieve the same messages. The disadvantage is that you end up retrieving mail you've already read and possibly replied to.

You can control whether messages are removed after they're picked up. This setting applies only to POP3 email, not to corporate (Exchange Server), Gmail, or IMAP (webmail) accounts. Follow these steps:

1. **Open the Settings app and choose Accounts.**

2. **Tap the Personal (POP3) item.**

   This item might be titled Email.

3. **Choose Account Settings.**

   The action might be titled Settings.

4. **Choose an email account.**

5. **Choose Server Settings or Incoming Settings**

   Look for a MORE SETTINGS button if you don't directly see either item.

6. **Look for the action bar by the Delete Email from Server item.**

7. **Choose the When Deleted from Inbox item.**

8. **Tap the DONE button.**

When these steps don't work, use the Email app to change the setting: Tap the Action Overflow (or MORE button), and choose Settings. Continue with Step 4 from the list.

After you make or confirm this setting, messages you delete on the phone are also deleted from the mail server. That means the message won't be picked up again, not by the phone, another mobile device, nor any other computer that fetches email from that same account.

Setting the primary email account

When you have more than one email account, the Email app uses that main account — the default — to send messages. To change that primary mail account, follow these steps:

1. **Start the Email app.**

2. **Choose Combined View or Combined Inbox.**

   This option is chosen from an action bar or the navigation drawer.

3. **Tap the Action Overflow or the MORE button.**

4. **Choose Settings.**

5. **On the Email Settings screen, tap the Action Overflow or the MORE button and choose Set Default Account.**

6. **Choose the email account you want to use as the default for sending messages.**

You can still choose an account when you send an email message: Manipulate the From field when composing a new message as described earlier in this chapter. Choose the account you want to use, and the message is sent from that account.

Chapter 10

Out On the Web

When Tim Berners-Lee developed the World Wide Web back in 1990, he had no idea that people would one day use it on their cell phones. Nope, the web was designed to be viewed on a computer — specifically, one with a nice, roomy, high-resolution monitor and a full-size keyboard. Cell phones? They had teensy LED screens. Browsing the web on a cell phone would have been like viewing the Great Wall of China through a keyhole.

Well, okay: Viewing the web on your cell phone is kind of like viewing the Great Wall through a keyhole. The amazing thing is that you can do it in the first place. For some web pages, it's workable. For others, you have my handy advice in this chapter.

TIP

» If possible, activate the phone's Wi-Fi connection before you venture out on the web. Though you can use the mobile data connection, a Wi-Fi connection incurs no data usage charges.

» Many places you visit on the web can instead be accessed directly and more effectively by using specific apps. Facebook, Gmail, Twitter, YouTube, and other popular web destinations have apps that you may find are already installed on your phone or otherwise available for free from Google Play.

>> One thing you cannot do with your phone is view Flash animations, games, or videos. The phone's web browser disables the Flash plug-in, also known as Shockwave. I know of no way to circumvent this limitation.

The Phone's Web Browser App

All Android phones feature a web browsing app. The stock Android app is Google's own Chrome web browser. Your phone may use another web browser app, and it may be given a simple name such as Web, Browser, or Internet. Each of these web browser apps works in a similar way and offers comparable features.

Here's the secret: Pretty much every phone's web browser app is simply the Chrome web browser app in disguise. That's good, because it's the web browser app I write about in this chapter.

>> If your phone doesn't have the Chrome app, you can obtain it for free from Google Play. See Chapter 16.

>> An advantage of using Chrome is that your bookmarks, web history, and other features are shared between all devices you use on which Chrome is installed. So, if you use Chrome as your computer's web browser, it's logical to use Chrome on your phone as well.

>> The first time you fire up the web browser app on certain Samsung phones, you may see a registration page. *Registering,* or signing up for a Samsung account, is optional.

Behold the Web

Rare is the person these days who has had no experience with the World Wide Web. More common is someone who has used the web on a computer but has yet to taste the Internet waters on a mobile device. If that's you, consider this section your quick mobile-web orientation.

Surfing the web on your phone

When you first open the web browser app, you're taken to the home page. The Chrome app lacks a home page, so instead you see the last page you were viewing.

Or, if you're starting the app for the first time, you see Google's main page, illustrated in Figure 10-1.

Here are some handy tips for web browsing on your phone:

Incognito tab Address box Tabs Action Overflow Forward Favorite Refresh

FIGURE 10-1: The Chrome app beholds the Google home page.

>> Drag your finger across the touchscreen to pan the web page. You can pan up, down, left, and right, which is convenient when the web page is larger than the phone's screen.

>> Pinch the screen to zoom out, or spread two fingers to zoom in.

>> The page you see may be the mobile version, customized and designed for small-screen devices. To see the nonmobile version, tap the Action Overflow icon and choose Request Desktop Site. (Refer to the right side of Figure 10-1.)

» You can reorient the phone horizontally to read a web page in Landscape mode.

Visiting a web page

To visit a web page, heed these directions in the web browser app:

1. Tap the address box.

Refer to Figure 10-1 for the address box's location. If you don't see the address box, swipe your finger from the top of the screen downward.

2. Use the onscreen keyboard to type the address.

You can also type a search word, if you don't know the exact web page address.

3. Tap the Go button on the onscreen keyboard to search the web or visit a specific web page.

You can also tap links on a web page to follow that link. If you have trouble tapping a specific link, zoom in on the page. On some phones, you long-press the screen to display a magnification window, which makes it easier to tap links.

» The onscreen keyboard may change certain keys to make it easier to type a web page address. Look for a www (World Wide Web) or .com (dot-com) key.

» Long-press the .com key to see other top-level domains, such as .org and .net.

» To reload a web page, tap the Refresh icon. If you don't see that icon on the screen, tap the Action Overflow to find the Refresh or Reload action. Refreshing updates a website that changes often. Using the Refresh action can also reload a web page that may not have completely loaded the first time.

» To stop a web page from loading, tap the Cancel (X) icon that appears by the address box.

Browsing back and forth

To return to a previous web page, tap the Back navigation icon.

To go forward after going back, tap the web browser app's Forward icon. This icon is found on the Action Overflow, as illustrated in Figure 10-1, though some web browser apps may place it next to the address box.

To review web pages you've visited, tap the Action Overflow and choose History. In some web browser apps, look for the History item on the Bookmarks card: Tap the Action Overflow and choose Bookmarks.

>> See the later section "Clearing your web browser history" for information on purging items from the History list.

>> If you find yourself frequently clearing the web page history, consider using an incognito tab. See the later section "Going incognito."

Working with favorites (bookmarks)

You may know them as bookmarks, but on your phone, saved page addresses are called favorites. To make a web page one of your favorites, follow these steps:

1. Navigate to the web page.

Favorite web pages are those you plan on visiting frequently or want to return to later.

2. Tap the Action Overflow and choose the Star icon.

The Star (Favorite) icon is shown in the margin, as well as earlier, in Figure 10-1.

The bookmark is added instantly, saved in the Mobile Bookmarks folder, which is illustrated on the left in Figure 10-2. If you want to make changes, tap the Action Overflow (illustrated in the figure) and choose Edit to change information on the bookmark's card (on the left in Figure 10-2).

For example, I typically shorten the bookmark name, which is often longer and more detailed than I need.

You can also choose to move the bookmark to a specific folder: Tap the Folder item to choose a location other than Mobile Bookmarks. For example, choose Desktop Bookmarks to save the website in a location accessible on the desktop computer version of Chrome.

Tap the Back navigation icon when you're done editing the bookmark.

>> To visit a favorite website, in the web browser app tap the Action Overflow and choose Bookmarks. Tap a bookmark in the list.

>> If you don't see the bookmarks you want, tap the Side Menu icon on the Bookmarks screen and choose another source, such as Bookmarks Bar, which lists your desktop bookmarks.

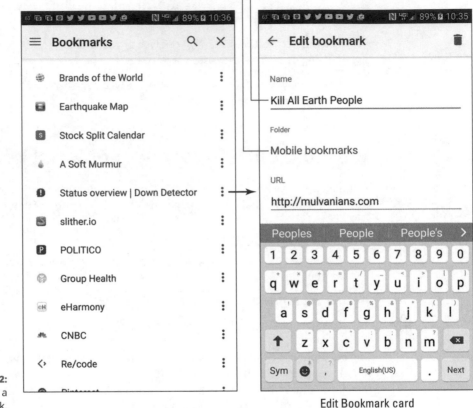

Set bookmark folder

Edit name

FIGURE 10-2:
Editing a
bookmark.

Edit Bookmark card

TIP

» To quickly visit a bookmarked website, just start typing the site's name in the address box. A list of matching results appears. Tap the bookmarked site from the list.

» To remove a bookmark, tap the Action Overflow and choose Bookmarks, and then tap the Action Overflow by a specific bookmark. Choose the Delete action. The bookmark disappears.

» A great way to find sites to bookmark is to view the web page history. See the preceding section.

Going incognito

TIP

Shhh! For private browsing, you can open an incognito tab: Tap the Action Overflow and choose New Incognito Tab. The Incognito tab takes over the screen, changing the look of the Chrome app and offering a description page.

When you go incognito, the web browser doesn't track your history, leave cookies, or provide other evidence of which web pages you've visited. This level of privacy doesn't prevent viruses or sophisticated snooping software.

When an incognito tab is open, the Incognito notification appears on the phone's status bar, similar to the one shown in the margin. Choose that notification to close all open incognito tabs.

See the next section for information on switching between incognito tabs and other tabs in the web browser app.

Managing multiple web pages in tabs

Like many popular web browsers, the Chrome app uses a tabbed interface to help you access more than one web page at a time. You can do various interesting things with tabs:

>> **To open a link in another tab,** long-press the link and choose Open in New Tab.

>> **To open a new tab,** tap the Action Overflow and choose New Tab.

>> **To switch tabs, tap the Tabs icon,** illustrated in Figure 10-3. Choose a new tab from the list shown.

The number of open tabs appears on the Tabs icon. This value doesn't include any incognito tabs, though they appear in the list of open tabs, as shown in Figure 10-3.

To close a tab, tap the Tabs icon and then tap the Close (X) icon by the tab thumbnail.

You can also close a tab by swiping it from the *Overview*, the list of recently opened apps.

Searching the web

The best way to find things on the web is to use the Google widget, found floating on the Home screen. Type your search item into that box, or utter "Okay, Google" and speak the search text.

Also see Chapter 15 for information on using Google Now.

Close tab

Add a new tab

Tabs

Incognito tab

FIGURE 10-3:
Switching tabs.

Choose a tab

Finding text on a web page

To locate text on a web page, tap the Action Overflow icon and choose Find in Page. Type the search text into the Find in Page box. As you type, found text is highlighted on the screen. Use the up or down chevrons to the right of the search box to page through the document.

Tap the Back icon to dismiss the Find in Page box after you've finished searching.

Sharing a web page

There it is! That web page of your dreams. You must tell everyone you know. The gauche way to share the page is to copy and paste it. Because you're reading this book, though, you know the better way to share a web page. Heed these steps:

1. **Go to the web page you desire to share.**

   Actually, you're sharing a *link* to the page, but don't let my obsession with specificity deter you.

2. **Tap the Action Overflow icon and choose Share.**

   The action might also be called Share Page or Share Via. Either way, a long list of apps appears. The variety and number of apps depends on what's installed on your phone.

3. **Choose an app.**

   For example, choose Gmail to send the link by email, or choose Facebook to share the link with your friends.

4. **Do whatever happens next.**

   Whatever happens next depends on how you're sharing the link: Compose the email, write a post in Facebook, or whatever. Refer to various chapters in this book for specific directions.

TIP

You cannot share a page from an Incognito tab.

Also see Chapter 18 for information on printing web pages.

The Joy of Downloading

Downloading is an easy concept to grasp, as long as you accept how most people misuse the term. Officially, a *download* is a transfer of information over a network from another source to your phone. For the web browser app, the network is the Internet, and the other source is a web page.

» The Download Complete notification appears after your phone has downloaded something. You can choose this notification to view the downloaded item.

» If you're quick, you can tap the OPEN button that appears on the *toast* (pop-up message) immediately after the download is complete.

» You use the Play Store app to *install* new apps on your phone. That's a type of downloading, but it's not the same as the downloading described in this section. See Chapter 16 for details.

TECHNICAL STUFF

>> Most people use the term *download* when they really mean *transfer* or *copy*. Those people must be shunned.

>> The opposite of downloading is *uploading*. That's the process of sending information from your gizmo to another location on a network.

Grabbing an image from a web page

The simplest thing to download is an image from a web page. Obey these steps:

1. Long-press the image.

You see an action card with a list of things you can do.

2. Choose Save Image.

The image is downloaded and stored on your phone.

See the later section "Reviewing your downloads" for details on how to access the image. It also appears in the Photos app, covered in Chapter 13.

Downloading a file

The web is full of links that don't open in a web browser window. For example, some links automatically download, such as links to PDF files or Microsoft Word documents or other types of files that a web browser is too a-feared to display.

To save other types of links that aren't automatically downloaded, long-press the link and choose the Save Link action. If this action doesn't appear, your phone is unable to save the link, because either the file is of an unrecognized type or it presents a security issue.

Reviewing your downloads

To review items downloaded from the Internet, choose the Download Complete notification, similar to the one shown in the margin.

When the Download Complete notification goes away, you can open the Downloads app to access all downloaded material. You'll find it eagerly awaiting your attention in the Apps drawer.

The Downloads app lists downloaded items, organized by date. To view a download, tap an item in the list. The phone then starts the appropriate app to view the item.

>> Some downloads you cannot view. That's because not every file is accessible on a mobile device.

>> You might be prompted to choose an app to open the document. See Chapter 19 for details on selecting default apps.

>> Downloaded material is saved to the phone's internal storage. It can be found in the Download or Downloads folder. Read about file storage in Chapter 18.

TECHNICAL
STUFF

Web Controls and Settings

More options and settings and controls exist for the web browser app than just about any other Android app. Rather than bore you with every dang-doodle detail, I thought I'd present just a few of the options worthy of your attention.

Clearing your web browser history

When you don't want the entire Internet to know what you're looking at on the web, open an incognito tab, as described earlier, in the section "Going incognito." When you forget to do that, follow these steps to clear one or more web pages from the browser history:

1. **Tap the Action Overflow and choose History.**

2. **Tap the X icon next to the web page entry you want to remove.**

   It's gone.

Changing the way the web looks

You can do a few things to improve the way the web looks on your phone. The key is to find the actions that adjust the text size or zoom level. Heed these steps:

1. **Tap the Action Overflow and choose Settings.**

2. **Choose Accessibility.**

3. **Use the Text Scaling slider to adjust the text size.**

   The preview text below the slider helps you gauge which size works best.

You can spread your fingers to zoom in on any web page. You can also turn the phone sideways to view a web page larger in the horizontal orientation.

REMEMBER

Setting privacy and security options

Web browser app security settings are most likely preset to a comfortable level of protection. One item you might consider checking is how information is retained and automatically recalled for the web pages you visit. You may want to disable some of those features. Obey these steps:

1. **Tap the Action Overflow and choose Settings.**

2. **Choose Privacy.**

3. **Tap the Clear Browsing Data button.**

   This step might not be necessary for all web browser apps, which directly show you the list of selected items.

4. **Place check marks by those items you want removed from the phone's storage.**

5. **Tap the CLEAR DATA button.**

Some web browser apps might present options for remembering form data or passwords. If so, remove the check marks by those items. Also, delete any personal data, if that's an option.

TIP

With regard to general online security, my advice is to be smart on the web and think before doing anything questionable. Use common sense. The most effective tool that the Bad Guys have is called *human engineering*. They use ways to trick you into doing something you normally wouldn't do, such as click a link to see a cute animation or a racy picture of a celebrity or politician. As long as you think before you click, you should be safe.

Also see Chapter 21 for information on applying a secure screen lock, which I highly recommend.

Chapter 11

Digitally Social

S ocial networking is a 21st century phenomenon that proves many odd beliefs about people. For example, it's possible to have hundreds of friends and never leave your house. You can jealously guard your privacy against the wicked intrusions of the government, all while letting everyone on the Internet know that you've just "checked in" to the dentist to have a root canal. And you can share your most intimate moments with humanity, many of whom will "like" the fact that you've just broken up or that your cat was run over by the garbage collection service.

Share Your Life on Facebook

Of all the social networking opportunities, Facebook is the king. It's the online place to go to catch up with friends, send messages, express your thoughts, share pictures and videos, play games, and waste more time than you ever thought you had.

>> The best way to access Facebook is to use the Facebook app. This app is preinstalled on some phones. If it isn't, you can obtain the Facebook app for free at Google Play. See Chapter 16.

TIP

>> You can use the Facebook app to sign up for a Facebook account, or you can use your existing account.

>> After signing in to Facebook the first time, you have to perform configuration. I recommend choosing the option to synchronize Facebook with your phone's contacts.

>> The Facebook app is updated frequently. Visit my website to review any new information:

```
www.wambooli.com/help/android
```

Using Facebook on your phone

The Facebook app presents information on several tabs, as shown in Figure 11-1. The tab you'll probably use the most is the News Feed. Options for interacting with Facebook appear near the top of the News Feed, as illustrated in the figure.

The Facebook app continues to run until you either sign out or turn off the phone. To leave the app, tap the Home navigation icon. Or you can tap the Recent navigation icon to switch to another, running app.

To sign out of the Facebook app, tap the More icon (refer to Figure 11-1) and choose the Log Out action. (Swipe down the screen to locate that action.) Tap the LOG OUT button to confirm.

>> Refer to Chapter 19 for information on placing a Facebook launcher or widget on the Home screen.

>> Swipe the News Feed downward to instantly refresh its contents.

>> Use the Like, Comment, or Share buttons below a News Feed item to comment, like, or share something, respectively. You can see any existing comments only when you choose the Comment item.

>> The Facebook app generates notifications for new news items, mentions, chats, and so on. The Facebook notification icon looks similar to the one shown in the margin.

Setting your status

The primary thing you live for on Facebook, besides having more friends than anyone else, is to update your status. It's the best way to share your thoughts with the universe — far cheaper than skywriting and far less offensive than a robocall.

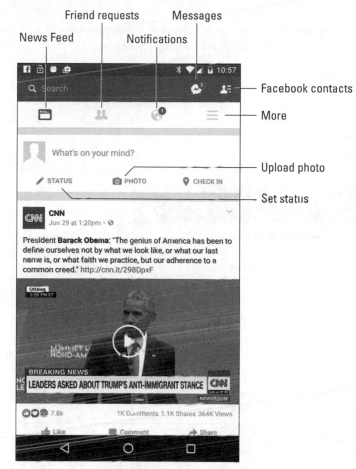

Friend requests   Messages

News Feed   Notifications

Facebook contacts

More

Upload photo

Set status

FIGURE 11-1:
The Facebook
app.

To set your status, follow these steps in the Facebook app:

1. **Switch to the News Feed.**

   Tap the News Feed icon, shown in Figure 11-1.

2. **Tap the status update area — specifically, the text *What's On Your Mind.***

   You see the Post to Facebook screen, where you can type your musings, similar to what's shown in Figure 11-2.

3. **Tap the sharing audience to choose where to send the new post.**

   Your choices are Public, so that everyone can see it; Friends, so that only people you're friends with can see it; and others. Public is the most popular.

4. **Tap the Back navigation icon to return to the post.**

5. Tap the *What's On Your Mind* text to type something pithy, newsworthy, or typical of the stuff you read on Facebook.

6. Tap the POST button to share your thoughts.

To cancel the post, tap the Back navigation icon. Tap the DISCARD POST button to confirm.

Choose sharing audience

FIGURE 11-2:
Updating your
Facebook
status.

Uploading a picture to Facebook

One of the many things your Android phone can do is take pictures. Combine that feature with the Facebook app, and you have an all-in-one gizmo designed for sharing the various intimate and private moments of your life with the ogling Internet throngs.

To start the picture-posting process, tap the Photo icon in the Facebook app. Refer to Figures 11-1 and 11-2 for popular Photo icon locations on the main screen and the Post to Facebook screen. After you tap the Photo icon, the photo selection screen appears. You have several choices:

>> First, you can select an image from pictures shown on the screen. These images are stored on the phone. Tap an image, or tap several images to select a bunch, and then tap the DONE button.

>> Second, you can take a picture and send it to Facebook. Tap the Add Photo icon, as shown in the margin, to begin the process.

>> Third, tap the Add Video icon to use the phone's camera and record a video.

If you elect to use the phone's camera to take a picture or shoot a video, you're switched to the Camera app. Snap the photo or record the video. You see a preview screen, similar to the one shown in Figure 11-3.

Cancel

FIGURE 11-3:
Recording video for Facebook.

Retry   Done/OK   Play
(video only)

Tap the Retry button to take another image, or tap Done / OK and get ready to post the image or video to Facebook. Tap Cancel (in the upper left corner of the screen) to abandon your efforts.

After you select the image or video, it appears on a new post screen. Continue to create the post as described earlier in this chapter. Tap the POST button to present your efforts to the Facebook community. The image can be found as part of your status update or News Feed, and it's also saved to your Mobile Uploads album on Facebook.

>> See Chapter 13 for more information on using the phone's camera.

>> I find it easier to use the Camera app to take a bunch of images and then choose an image later to upload it to Facebook.

>> To share something on Facebook when you use another app, tap the Share icon. Choose the Facebook app to share what you're viewing: an image, music, YouTube video, and so on.

Changing the Facebook ringtone

To change or eliminate the noise the Facebook app makes for notifications, follow these steps:

1. **Tap the More icon.**

   Refer to Figure 11-1 for its location.

2. **Choose App Settings.**

3. **Choose Notifications.**

4. **Choose Notification Ringtone.**

5. **If prompted by the Complete Action Using card, select Media Storage and tap the JUST ONCE button.**

   See Chapter 19 for information on dealing with the Complete Action Using card, which falls under the topic of "default apps."

6. **Choose a sound from the list.**

   You hear a sound preview, but nothing is set until you:

7. **Tap OK.**

To mute Facebook notifications, choose the sound None in Step 6.

To use the phone's standard notification ringtone, choose Default Notification Sound in Step 6.

Also see Chapter 20 for information on setting the phone's ringtone.

A Virtual Hangout

Google has made various forays into the social networking arena, the most obvious of which is the Google+ service. Though the Google+ app can be found on your phone, a better and more useful social app is Google Hangouts.

The Hangouts app offers text, voice, and video chat. On some phones, Hangouts serves as the text messaging app; see Chapter 8. The Hangouts app can even be used to place phone calls, as long as the companion app, Hangouts Dialer, is installed.

Using Hangouts

Start your Google Hangouts adventure by opening the Hangouts app. You may find a launcher on the Home screen, the app might dwell in a Google folder, or, like all apps on the phone, it can be found in the Apps drawer.

When you first start the Hangouts app, it may ask whether you want to make phone calls. Of course you do! Install the Hangouts Dialer app. If you're not prompted, get that app from Google Play.

Hangouts hooks into your Google account. You can connect with any one of your contacts who also shares a Google or Gmail account. Obey these steps:

1. Tap the Side Menu icon.
2. Choose Contacts from the navigation drawer.
3. Tap a contact.
4. Choose whether to text-chat (Hangouts Message), voice-call, or video-chat.
5. Connect with the other person.

If you choose Hangouts Message, you see a screen similar to the text messaging app on your phone. Type a message to the other person. That person types back (or not), and so it goes.

In Figure 11-4, you see the main Hangouts app screen. It shows previous conversations and indicates the type of chat: text (no icon), video, or voice.

You can do anything else on the phone while the Hangouts app stays active. You're alerted via notification of an impending Hangouts request. The notification icon is shown in the margin.

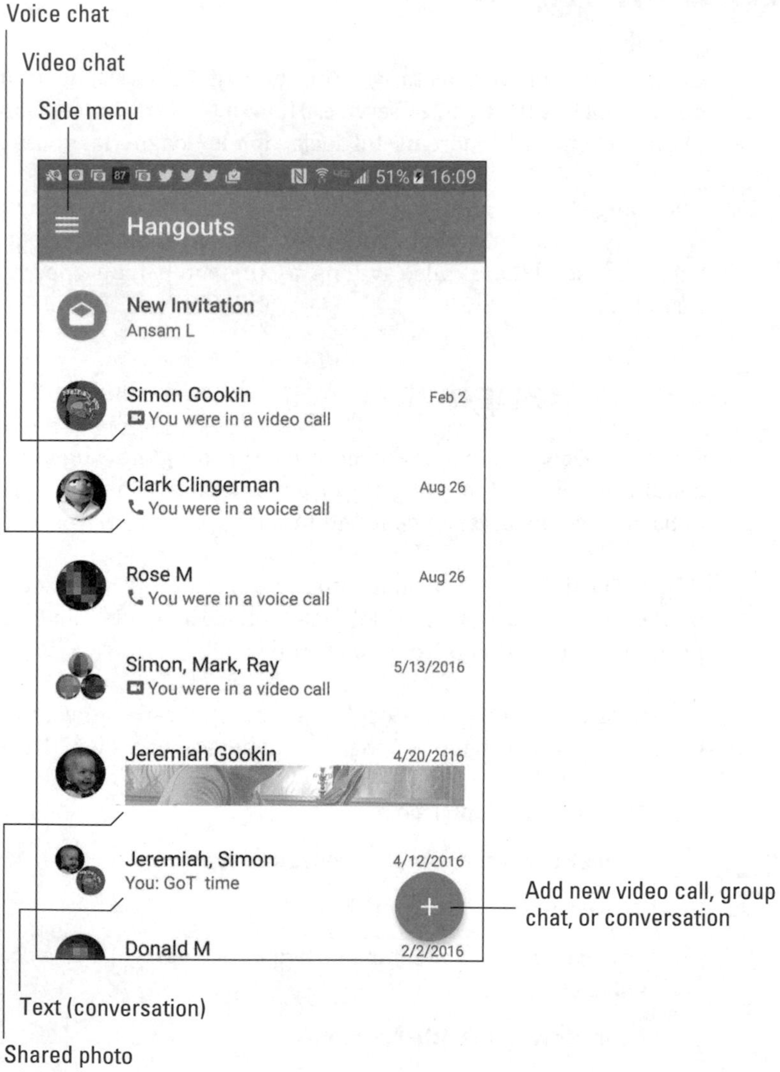

Voice chat

Video chat

Side menu

	87	N 📶 51% 🔋 16:09
≡	**Hangouts**	

✉️ **New Invitation**
Ansam L

Simon Gookin   Feb 2
📹 You were in a video call

Clark Clingerman   Aug 26
📞 You were in a voice call

Rose M   Aug 26
📞 You were in a voice call

Simon, Mark, Ray   5/13/2016
📹 You were in a video call

Jeremiah Gookin   4/20/2016

Jeremiah, Simon   4/12/2016
You: GoT time

Donald M   2/2/2016

Add new video call, group
chat, or conversation

FIGURE 11-4:
Google
Hangouts.

Text (conversation)

Shared photo

>> Conversations are archived in the Hangouts app. To peruse a previous text chat, tap its entry on the main screen. (Refer to Figure 11-4.)

>> Video calls aren't archived, but you can review when the call took place and with whom by choosing a video chat item.

>> To remove a previous conversation, long-press it. Tap the Trash icon that appears atop the screen.

>> When installed, the Hangouts Dialer app lets you place phone calls from within the Hangouts app. The Dialer tab appears atop the main Hangouts app screen (not shown in Figure 11-4). Tap that tab, and then type a contact name or phone number. Tap the matching contact or phone number to dial.

REMEMBER

>> To use Hangouts, your friends must have Google accounts. They can be using computers or mobile devices; it doesn't matter which. But each person must have a camera available to enable video chat.

Typing at your friends

Text chatting is one of the oldest forms of communication on the Internet. People type text back and forth at each other, which can be tedious, but it remains popular. To text-chat in the Hangouts app, obey these steps:

1. **Choose a contact or select a previous conversation from the main Hangouts app screen.**

2. **Use the onscreen keyboard to type a message, as shown in Figure 11-5.**

3. **Tap the Send icon to send your text tidbit.**

You type, your friend types, and so on until you grow tired or the phone's battery dies.

>> To add more people to the hangout, tap the Action Overflow and choose New Group Conversation. Choose more friends from those listed to invite them into the hangout.

>> Tap the Attachment icon to add a photo to the conversation, such as the image shown in Figure 11-5. The Attachment icon, shown in the margin, appears in the Send Hangouts Message text box, but only when you haven't yet typed anything.

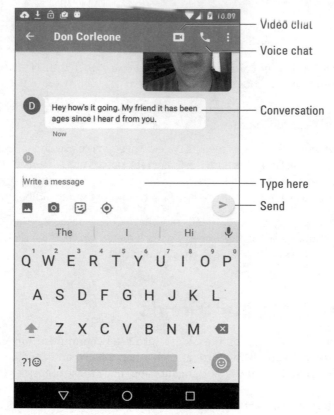

Video chat

Voice chat

Conversation

Type here

Send

FIGURE 11-5:
Text chatting.

Talking and video chat

Take the hangout up a notch by tapping the Voice Chat icon. (Refer to Figure 11-5.) When you do, your friend receives a pop-up invite. After that person taps the Accept icon, you begin talking.

To see the other person, tap the Video Chat icon. As soon as they agree, you can see each other, as shown in Figure 11-6. The person you're talking with appears in the big window; you're in the smaller window, as shown in the leftmost figure.

To mute the call, tap the Microphone icon. In Figure 11-6, the microphone is muted. Tap the Video icon (it's in the lower right of the right side of Figure 11-6) to disable the phone's camera and return to voice or text chat.

The onscreen controls may vanish after a second; tap the screen to see the controls again.

To end the conversation, tap the End Call icon, at the bottom center. Well, say goodbye first, and then tap the icon.

Switch cameras

Privacy

Person you're calling

You

FIGURE 11-6:
A video chat.

Mute   End   Disable
        call   camera

Let's All Tweet

Twitter is a social networking site that lets you share short bursts of text, or *tweets.* You can create your own or simply choose to follow others, including news organizations, businesses, governments, celebrities, and creatures from alien planets.

If your phone didn't come with the Twitter app preinstalled, obtain it from Google Play, as described in Chapter 16. Install and run the app to sign in to Twitter using an existing account, or create a new account on the spot.

Figure 11-7 shows the Twitter app's main screen, which shows the current tweet feed. The Twitter app is updated frequently, so its exact appearance may change after this book has gone to press.

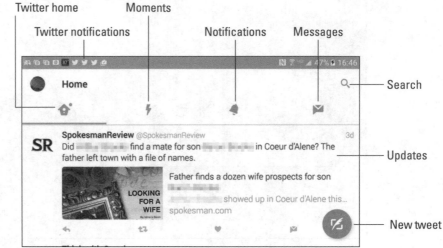

Twitter home

Moments

Twitter notifications

Notifications

Messages

Search

Updates

New tweet

FIGURE 11-7:
The Twitter app.

To read tweets, choose the Home category, shown in Figure 11-7. Recent tweets are displayed in a list, with the most recent information at the top. Scroll the list by swiping it with your finger. Drag the list downward to refresh.

To tweet, tap the New Tweet icon, shown in Figure 11-7. The new tweet screen appears, where you can compose your tweet.

REMEMBER

A tweet can be no more than 140 characters long. That number includes spaces and punctuation. For direct messages, Twitter has raised the limit above 140 characters.

Tap the Tweet button to share your thoughts with the twitterverse.

TIP

» The Twitter app comes with companion widgets that you can affix to the Home screen. Use the widgets to peruse recent tweets or compose a new tweet. Refer to Chapter 19 for information on affixing widgets to the Home screen.

» You can share material from your phone with Twitter by tapping the Share icon in various apps. The content you're looking at is then added to a new tweet.

Skype the World

The Skype service is used the world over as a way to make free Internet phone calls and to video-chat. Plus, if you're willing to pony up some money, you can make inexpensive international phone calls.

The typical Android phone doesn't come with the Skype app preinstalled. To get Skype, visit Google Play and obtain the Skype app.

To use Skype, you need a Skype account. You can sign up when you first open the app, or you can visit www.skype.com on a computer and complete the process by using a nice, full-size keyboard and widescreen monitor.

Skype's popular features include text chat as well as voice and video chat, although you can use these features only with fellow Skype users. The big enchilada in Skype is placing phone calls. Because you already have a phone, that feature's most important aspect is placing cheap international calls. Here's how it works:

1. **Tap the Dialpad icon on the Skype app's Home screen.**

   The Dialpad icon is shown in the margin. After you tap this icon, you see the Skype dial screen, illustrated in Figure 11-8.

2. **Use the keypad to type the international number.**

   The number begins with a plus sign (+), followed by the country code and then the phone number. You can choose a country by tapping the screen, as shown in Figure 11-8.

3. **Tap the Dial icon to dial the number.**

4. **When the other party answers, talk.**

These steps meet with success only when you've purchased Skype Credit. If you haven't purchased it, tap the Side Menu icon in the Skype app and choose Skype Credit. Don't worry about breaking the bank! You don't need a lot of Skype Credit to place an international call.

>> Text, voice, and video chat on Skype over the Internet are free. When you use a Wi-Fi connection, you can chat without consuming your cellular plan's data minutes.

>> If you plan to use Skype a lot, get a good headset.

TIP

Select country code   Backspace/Delete

FIGURE 11-8:
The Skype app
dialing screen.

Dial

3 Amazing Phone Feats

Chapter 12

There's a Map for That

Where are you? More importantly, where is the nearest Mexican restaurant? Normally, these questions can be best answered when you're somewhere familiar or with someone who knows the territory or while standing at a brightly lit intersection and the aroma of chili peppers and tequila is wafting from a restaurant nearby and the sign says *El Viejo Buey*.

The answer to the questions "Where are you?" and "How to get somewhere else?" are readily provided by the Maps app. This app uses your phone's global positioning system (GPS) capabilities to gather information about your location. It also uses Google's vast database to locate interesting places nearby and even get you to those locations. The Maps app is truly amazing and useful — after you understand how it all works.

Basic Map

Perhaps the best thing about using the Maps app is that there's no risk of improperly folding anything. Even better, the Maps app charts everything you need to find your way: freeways, highways, roads, streets, avenues, drives, bike paths, addresses, businesses, and points of interest.

Using the Maps app

Start the Maps app by tapping its launcher on the Home screen, or you can summon the app from the Apps drawer. If you're starting the app for the first time or it has recently been updated, you must agree to the terms and conditions. Do so as directed on the touchscreen.

Your Android phone communicates with GPS satellites to hone in on your current position on Planet Earth. That location appears on the map, similar to Figure 12-1. The position is accurate to within a given range, as shown by a blue circle around your location. If the circle doesn't appear, your location is either pretty darn accurate or you need to zoom in.

Here are some fun things you can do when viewing the basic street map:

Zoom in: To make the map larger (to move it closer), double-tap the screen. You can also spread your fingers on the touchscreen to zoom in.

Zoom out: To make the map smaller (to see more), pinch your fingers on the touchscreen.

Pan and scroll: To see what's to the left or right or at the top or bottom of the map, swipe your finger on the touchscreen. The map scrolls in the direction you swipe.

Rotate: Using two fingers, rotate the map clockwise or counterclockwise. Tap the Compass Pointer icon, as shown in the margin, to reorient the map with north at the top of the screen.

Perspective: Touch the screen with two fingers and swipe up or down to view the map in perspective. You can also tap the Location icon to switch to perspective view, although that trick works only for your current location. To return to flat-map view, tap the Compass Pointer icon.

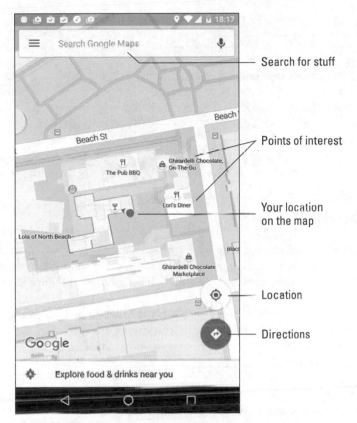

Search for stuff

Points of interest

Your location
on the map

Location

Directions

FIGURE 12-1:
Your location
on the map.

The closer you zoom in to the map, the more detail you see, such as street names, address block numbers, and businesses and other sites — but no tiny people.

» See the nearby sidebar "Activate the location technologies" to confirm that the phone is presenting your location accurately.

» The blue triangle (refer to Figure 12-1) shows in which general direction the phone is pointing.

» When the phone's direction is unavailable, you see a blue dot as your location on the map.

» The Location icon changes to the Perspective icon when the Maps app is in perspective view. Tap the Perspective icon to return to flat-map view.

» When all you want is a virtual compass, similar to the one you lost as a kid, get a compass app from Google Play. See Chapter 16 for more information about Google Play.

ACTIVATE THE LOCATION TECHNOLOGIES

The Maps app works best when you activate all the phone's location technologies, including both GPS and Wi-Fi. To ensure that these technologies are in use, open the Settings app and choose the Location item. On some Samsung phones, the Location item is found on the Connections tab. Ensure that the master control is in the On position. Further, if a Mode or Locating Method item is available, choose it and select the High Accuracy option.

Adding layers

You add details to the map by applying layers: A layer enhances the map's visual appearance by providing more information, or by adding other fun features to the basic street map, such as the Satellite layer, shown in Figure 12-2.

The key to accessing layers is to tap the Side Menu icon to view the navigation drawer. It displays several layers you can add to the map, such as the Satellite layer, shown in Figure 12-2. Another popular layer is Traffic, which lists updated travel conditions.

To remove a layer, choose it again from the navigation drawer; any active layer appears highlighted. When a layer isn't applied, the street view appears.

Saving an offline map

For times when an Internet connection isn't available (which is rare for an Android phone), you can still use the Maps app, though only in a limited capacity. The secret is to save the portion of the map you need to reference. Obey these steps while using the Maps app and when the phone has an Internet connection:

1. **View the map chunk you desire to save.**

   Zoom. Pan. Square in the area to save on the screen. It can be as large or as small as you need. Obviously, smaller maps occupy less storage.

2. **Tap the Side Menu icon.**

3. **Choose Offline Areas from the navigation drawer.**

   Any maps you've previously saved appear in the list.

4. **Tap the Add icon.**

5. **Tap the DOWNLOAD button.**

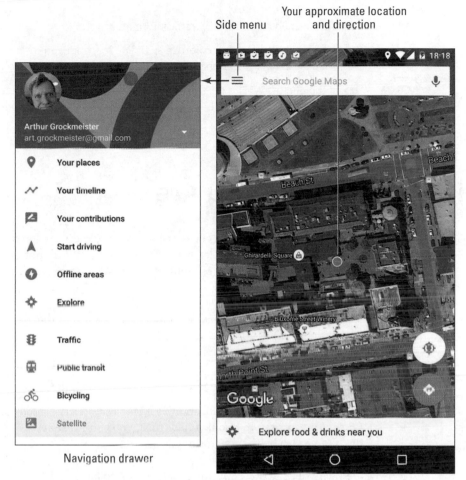

Side menu

Your approximate location and direction

Arthur Grockmeister
art.grockmeister@gmail.com

Search Google Maps

9 ▼◢ 🔋 18:18

Beach St

Beach St

Ghirardelli Square

Buxome Street Winery

North Point St

Google

Explore food & drinks near you

📍	Your places	
〰	Your timeline	
🖉	Your contributions	
▲	Start driving	
⊙	Offline areas	
✦	Explore	
🚦	Traffic	
🚇	Public transit	
⚲	Bicycling	
	Satellite	

FIGURE 12-2:
The Satellite
layer.

Navigation drawer

6. **Type a title for the map.**

Or you can keep the preset title — if the name makes sense to you.

7. **Tap the SAVE button.**

The map appears in the list of offline maps.

To use an offline map, display the navigation drawer and choose Offline Areas. Tap the offline map to view, and it shows up on the screen whether an Internet connection is active or not.

>> Offline maps remain valid for 30 days. After that time, you must update the map to keep it current.

>> To update an offline map, choose it from the Offline Areas screen and tap the UPDATE button.

>> Offline maps don't display traffic information.

>> You can zoom and pan to peruse an offline map, but you cannot search the map. Searching works only when an Internet connection is active.

>> If the phone's Wi-Fi is on, the offline map may display updated location information. This feature works even when the phone isn't connected to a specific Wi-Fi network.

It Knows Where You Are

Many war movies have this cliché scene: Some soldiers are looking at a map. They wonder where they are, when one of them says, "We're not even on the map!" Such things never happen on your phone's Maps app. That's because it always knows where you are.

Well, unless you're on the planet Venus. I've heard that the Maps app won't work there.

Finding a location

The Maps app shows your location as a blue dot on the screen. But *where* is that? I mean, if you need to phone a tow truck, you can't just say, "I'm the blue dot on the orange slab by the green thing."

Well, you *can* say that, but it probably won't do any good.

To view your current location, tap the Location icon, as shown in the margin. If you desire more information, or if you want details about any random place, long-press the Maps screen. Up pops a card, similar to the one shown on the left in Figure 12-3. The card gives your approximate address.

Tap the card to see a screen with more details and additional information, as shown on the right in Figure 12-3.

>> This trick works only when Internet access is available and the Maps app can communicate with the Google map servers.

>> To make the card go away, tap anywhere else on the map.

Long-press a location to see the address

Route

Mark the location as a favorite

Share this location

37.799452,-122.441025

3201-3213 Scott St
San Francisco, CA 94123

6 min

SAVE   LABEL   SHARE

Explore food & drinks here

Report a problem on 3201-3213 Scott St

FIGURE 12-3:
Finding an address.

Information card

Street view

>> For nearby locations, the Route icon (appearing as a bicycle in Figure 12-3) shows a method of travel and an approximate time value for the trip duration. Also see the later section "Your Phone Is Your Copilot."

TIP

>> When you have way too much time on your hands, play with the Street View feature. Tap that item on the location's card to examine a location preview with a 360-degree perspective. It's a great way to familiarize yourself with a destination or to plan a burglary.

Helping others find your location

TIP

You can use the Maps app to send your current location to a friend. If your pal has a phone with smarts similar to those of your Android phone, he can use the coordinates to get directions to your location. Maybe he'll bring tacos!

To send your current location to someone else, obey these steps:

REMEMBER

1. **Long-press your current location on the map.**

   To see your current location, tap the Location icon in the lower right corner of the Maps app screen.

   After long-pressing your location (or any location), you see a card displayed showing the approximate address.

2. **Tap the card and then tap the Share icon.**

   Refer to Figure 12-3 for this icon's location.

3. **Choose the app to share the location.**

   For example, choose the phone's text messaging app, Gmail, Email, or whichever useful app is listed.

4. **Use the selected app to complete the process of sending your location to someone else.**

As an example, you can use the phone's text messaging app to share your location. The recipient taps the link in the text message to open your location in his phone's Maps app. When the location appears, the recipient can follow my advice in the later section "Your Phone Is Your Copilot" to reach your location. Don't loan him this book, either — have him purchase his own copy. And bring tacos. Thanks.

Find Things

The Maps app can help you find places in the real world, just as the Google Search app helps you find places on the Internet. Both operations work similarly: Open the Maps app and type something to find into the Search text box. (Refer to Figure 12-1.) You can type a variety of terms in the Search box, as explained in this section.

Looking for a specific address

To locate an address, type it in the Search box; for example:

```
1600 Pennsylvania Ave., Washington, DC 20006
```

As you type, a list of suggestions appears. Tap a matching suggestion to view that location. Otherwise, tap the onscreen keyboard's Search key, and that location is shown on the map.

After finding a specific address, the next step is getting directions. See the later section "Navigating to your destination."

>> If you omit the city name or zip code, the Maps app looks for the closest matching address near your current location.

>> Tap the X button in the Search box to clear a previous search.

Finding a business, restaurant, or point of interest

You may not know an address, but you know when you crave sushi or perhaps the exotic flavors of Kamchatka. Maybe you need a hotel or a gas station or you have to find a place that patches toupees. To locate a business entity or a point of interest, type its name in the Search box; for example:

```
movie theater
```

This search text locates movie theaters on the current Maps screen. Or, to find locations near you, first tap the Location icon (shown in the margin) and then type the search text.

To look for points of interest at a specific location, add the city name, district, or zip code to the search text. For example:

```
sushi 98109
```

After typing this command and tapping the Search button, you see a smattering of sushi bars found near downtown Seattle, similar to those shown on the left side of Figure 12-4.

To see more information about a result, tap its card, such as the one for I Love Sushi on Lake Union in Figure 12-4. To get to the location, tap the Route icon on the location's card, as illustrated in the figure. See the later section "Navigating to your destination."

>> Every dot on the screen represents a search result. (Refer to Figure 12-4.)

>> Spread your fingers on the touchscreen to zoom in to the map.

>> The search results card (on the right in Figure 12-4) lists business hours as well as other information. Tap the Call icon to place a call. Tap the Website icon to visit the location's website, if available.

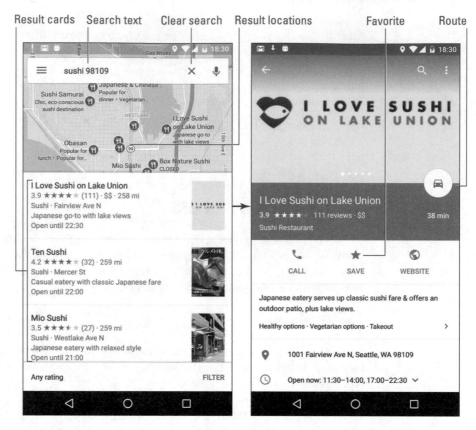

Result cards Search text Clear search Result locations Favorite Route

FIGURE 12-4:
Search results
for sushi in
Seattle.

TIP

>> If you really like the location, tap the SAVE (Star) icon. That action directs the Maps app to keep the location as one of your favorite places. The location is flagged with the Star icon on the Maps app screen. See the next section.

>> When you can't make up your mind or you just want to see what's out there, tap the Location icon and choose the item Explore Food & Drinks, found at the bottom of the screen. (Refer to Figure 12-1.) Peruse the list of locations to find something interesting.

Searching for favorite or recent places

Just as you can bookmark favorite websites on the Internet, you can use the Maps app to mark favorite places in the real world. The feature is called Your Places.

To review your favorite places or browse your recent map searches, tap the Side Menu icon and choose Your Places from the navigation drawer. The Your Places screen features four tabs:

>> **Labeled** lists locations related to your contacts, as well as your home and work locations. (See the later section "Setting your Home and Work locations.")

>> **Saved** lists any locations you've flagged as favorites. Tap the SAVE (Star) icon on the location's details card to save a location.

>> **Visited** refers to any locations you've been to recently.

>> **Maps** lets you view custom maps you've created. It's a feature not yet fully implemented as this book goes to press.

Select a tab, and then swipe through the list to see recent searches, saved places, and any offline maps you've saved. To revisit a location or view an item, tap its entry in the list.

Your Phone Is Your Copilot

The real point of having a map and finding a location is to get somewhere else. In the old days, you'd use your eyeballs to plot your route or rely upon directions from an acquaintance or a friendly local. With your Android phone, you tap the Route icon and you have a list of directions handy. The phone can even guide you on your way.

Setting your Home and Work locations

Two places you visit most frequently in the real world are where you live and where you work. You can preset these locations in the Maps app, which helps with navigation features. Follow these directions:

1. **In the Maps app, tap the Side Menu icon to display the navigation drawer.**

2. **Choose Your Places.**

3. **Tap Home.**

4. **Type your home address.**

   As you type, matching addresses appear in a list. As a shortcut, select one that represents your home location.

5. **Tap Work.**

6. **Type your work address.**

   Or you can set a place you frequent, such as your social club, church, or favorite pub.

7. **When the two locations are set, tap the Back navigation icon to return to the Maps app.**

The Home and Work addresses work as shortcuts when you use the Maps app to get directions. That topic is covered next.

>> If you choose Home or Work from the Your Places menu, you see that location displayed in the Maps app.

>> You can type **Home** or **Work** in the Maps app search box to instantly view those locations.

>> To update your home or work locations, display the Your Places screen (refer to Steps 1 and 2), but tap the Action Overflow icon by either location and choose the Edit action.

Navigating to your destination

One command that's associated with locations found in the Maps app deals with getting directions. The command is called Route, and it shows either the Route icon (see the margin) or a mode of transportation, such as car, bike, bus, or orni-thopter. Here's how it works:

1. **Tap the Route icon on a location's card.**

   You see one or more routes illustrated on the screen, similar to what's shown in Figure 12-5. The top of the screen shows either your current location, "Your location," or a specific address. The bottom location shows the destination, such as the I Love Sushi restaurant in the figure.

2. **If necessary, set a starting point.**

   The starting point is listed as Your Location, which is the phone's current location. You can type another location or use the Home or Work shortcuts, as described in the preceding section.

   If the starting point and the destination are reversed, tap the Action Overflow and choose Swap.

TIP

Starting location

Mode of transportation

Starting point

Destination

Navigation

Route card

Destination

FIGURE 12-5:
FIGURE 12-5:
Going some-
where else.

3. Choose a method of transportation.

The available options vary, depending on your location. In Figure 12-5, the items are (from left to right) Car, Public Transportation, On Foot, Ride Service, and Bicycle.

The route card's information is updated when you change the transportation method.

4. **To select another route, tap it on the screen.**

   Alternate routes appear gray. (Refer to Figure 12-5.) When you choose one, check the route card to observe time and distance differences.

5. **To view a list of directions, tap the route card.**

   You see turn-by-turn instructions for getting to your destination.

To begin turn-by-turn Navigation mode, tap the Navigation icon, as shown in Figure 12-5 and in the margin. You can mute the voice by tapping the Action Overflow and choosing Mute. Otherwise, toodle on to your destination.

To exit from Navigation mode, tap the Close icon on the screen.

>> Two things are critical when you use your phone to navigate while driving. First, plug the phone into a power source, such as the car's 12-volt power supply, formerly known as a cigarette lighter. Second, obtain a windshield-mount phone cradle. Together, these items can help you quickly and safely get to your destination.

>> The map shows your route, highlighted as a blue line on the screen. Areas of increased traffic are shown as orange (slow) or red (stopped). Road construction is also shown on the route, as are any toll roads or bridges.

TIP

>> If you don't like the route, you can adjust it: Drag the colored route line by using your finger. Time-and-distance measurements shown on the cards change as you adjust the route.

>> The Ride Service method of transportation isn't available everywhere. When it appears as an option, you see a card linking you to the Uber service, which you can use to get to your destination.

>> You may not get perfect directions from the Maps app, but it's a useful tool for places you've never visited.

REMEMBER

>> The phone stays in Navigation mode until you exit. A navigation notification can be seen atop the touchscreen while you're in Navigation mode.

Chapter 13

Pics and Vids

The second thing that Alexander Graham Bell thought of, right after the busy signal, was the camera phone. Sadly, it proved somewhat impractical because cameras and phones of the 1890s weren't really portable. And it was difficult to talk on the phone and hold still for 60 seconds to get the best exposure.

Not until the end of the 20th century did the marriage of the cell phone and digital camera become successful. It may seem like an odd combination, yet it's quite handy to have the phone act as a camera, because most people keep their phones with them. That way, should you chance upon Bigfoot or a UFO, you can always take a picture or video — even when there's no cell signal, which is often the case when I try to photograph Bigfoot.

The Phone's Camera

A camera snob will tell you, "No true camera has a ringtone." You know what? He's correct: Phones don't make the best cameras. Regardless, your phone has a camera. It can capture both still and moving images. You use the Camera app to carry out that task.

Here's the rub: Each phone seems to have its own Camera app. Basic controls are available but implemented differently in each app. Some Camera apps sport more controls, such as image effects, and other Camera apps are sparse.

This section uses the Google Camera app as an example. Your phone's Camera app may be identical or subtly different. And if you still can't figure out your phone's Camera app, you can obtain the Google Camera app from Google Play. See Chapter 16 for details on Google Play.

>> The Camera app features two basic modes: Still Shot and Video. The same app handles both shooting modes.

>> Many camera apps feature additional shooting modes, such as Panorama.

>> When you use the Camera app, the navigation icons (Back, Home, Recent) turn into tiny dots or vanish altogether. The icons are still there, and they still work; tap the screen or swipe from top to bottom to view the navigation icons.

Using basic camera controls

Here are some pointers that apply to all Android phone Camera apps:

>> Use the phone in either landscape or portrait orientation while taking a still shot. Don't worry either way: The image is always saved with the proper side up.

>> I strongly recommend that you record video in horizontal orientation only. This presentation appears more natural.

>> The phone's touchscreen serves as the viewfinder; what you see on the screen is exactly what appears in the final photo or video.

>> Tap the screen to focus on a specific object. You see a focus ring or square that confirms how the camera lens is focusing. Not every phone's camera hardware can focus; the front-facing camera features a fixed focus.

>> Zoom in by spreading your fingers on the screen.

>> Zoom out by pinching your fingers on the screen.

>> Some phones let you use the Volume key to zoom in or out: Up volume zooms in, Down volume zooms out. If the Volume key doesn't zoom, it may serve as a second shutter button (to snap the picture). While you're shooting video, the Volume key may serve to snap a still shot.

>> Hold the phone steady! I recommend using two hands for taking a still shot and shooting video.

TECHNICAL STUFF

» The phone stores pictures and videos in the DCIM/Camera folder. Still images are stored in the JPEG or PNG file format; video is stored in the MPEG-4 format. If your phone offers removable storage, the Camera app automatically saves images and videos to that media.

Snapping a still image

Taking a still image requires only two steps: First, ensure that the Camera app is in Still Shot mode; second, tap the Shutter icon to snap the photo.

In the Google Camera app, ensure that you're in single-shot mode by checking for the Shutter icon, shown in the margin. The screen should resemble Figure 13-1. Frame the image in the viewfinder, and then tap the Shutter icon to take the picture. You may hear a shutter sound, or the viewfinder may flash.

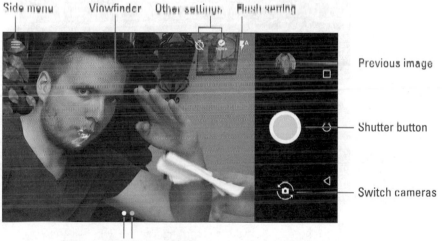

Side menu   Viewfinder   Other settings   Flash setting

Previous image

Shutter button

Switch cameras

FIGURE 13-1:
The Google
Camera app.

Swipe left for   Swipe right for
Still Shot mode   Video mode

Other camera apps sport features similar to those shown in Figure 13-1. You may see a switch or another icon to change from Still Shot mode to Video mode. The Shutter icon may look different. Different settings icons appear on the screen. No matter what the difference, tap the Shutter icon to take the picture.

REMEMBER

Set the resolution before you shoot. See the later section "Setting the resolution."

Recording video

To capture video, switch the Camera app to Video Recording mode. The Shutter icon becomes the Record icon, similar to the one shown in the margin. Tap that icon to start recording. The elapsed time and maybe even the amount of storage consumed appear on the touchscreen as video is being recorded, as shown in Figure 13-2. Tap the Stop icon to end the recording.

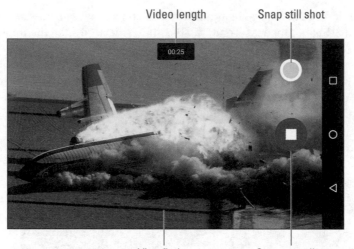

Video length   Snap still shot

00:25

Viewfinder   Stop recording

In the Google Camera app, swipe the screen to the right for Video mode, as shown in Figure 13-1. For other camera apps, tap the Video Mode icon to either switch modes to video or simply start recording.

>> Some Camera apps feature a Pause icon for recording video. Tap that icon to temporarily suspend recording. Tap the Record icon to resume. The video you record is saved as a single video as opposed to multiple videos.

>> Some versions of the Camera app may allow you to grab a still image while video is recording: Tap the screen. In Figure 13-2, you tap the Shutter icon to take a still image while recording.

REMEMBER

>> Hold the phone horizontally when you record video.

Deleting immediately after you shoot

Sometimes, you just can't wait to banish an image to bit hell. Either an annoyed person is standing next to you, begging that the photo be deleted, or you're just not happy and you feel the urge to smash the picture into digital shards. Hastily follow these steps:

1. **Tap the image thumbnail that appears on the screen.**

   The thumbnail is shown in Figure 13-1. If you don't see it, swipe the screen from right to left to see the previous image.

2. **Tap the Delete icon.**

   If you don't see the icon, tap the screen so that the icon shows up.

3. **If prompted, tap the OK or DELETE button to confirm.**

See the later section "Your Digital Photo Album" for more information on managing the phone's images.

Setting the flash

Camera apps feature three flash settings, illustrated in Table 13-1. The current setting appears on the screen (refer to Figure 13-1), or you may have to access the flash setting control to view the current setting.

TABLE 13-1 **Camera Flash Settings**

Setting	Icon	When the Flash Activates
Auto	⚡A	During low-light situations
On	⚡	Always
Off	⚡̸	Never

To change or check the flash setting in the Google Camera app, tap the current flash icon shown on the app's main screen. Tap that icon to cycle through and set the camera's flash setting. In other Camera apps, tap the Settings icon to access the Flash setting.

TIP

>> The Flash On setting works best when taking pictures of people or objects in front of something bright, such as a lovely calico kitten purring in front of an exploding volcano.

>> A "flash" setting is also available for shooting video in low-light situations. In that case, the flash LED is on the entire time. This setting is made similarly to setting the flash for still images, although the options are only On and Off. It must be set before you shoot video, and, yes, it devours a lot of battery power.

Setting the resolution

You don't always have to set the highest resolution or top quality for images and videos. Especially when you're shooting for the web or uploading pictures to Facebook, top quality is a waste of storage space because the image is shown on a relatively low-resolution computer monitor.

As you may suspect, setting the image resolution or video quality is done differently in each phone's Camera app. In the Google Camera app, follow these steps to access the resolution and video quality settings:

1. **Tap the Side Menu icon on the Camera app's view screen.**

You see the various shooting modes.

2. **Tap the Settings icon.**

3. **Choose Resolution & Quality.**

The Resolution & Quality screen is organized by shooting mode: Camera (still image) and Video. It's further organized by back or front camera.

4. **Choose a mode and a camera.**

For example, tap Back Camera Photo to set the still-image resolution for the phone's rear camera.

5. **Choose a resolution or video quality setting from the list.**

Options are presented in aspect ratio as well as in megapixels.

For other Camera apps, tap an Action Overflow or Settings icon and look for an action to set resolution or video quality. You may also have to switch camera modes or even between front and rear cameras before making the change.

REMEMBER

>> Set the resolution or video quality before you shoot! Especially when you know where the video will end up (on the Internet, on a TV, or in an email), it helps to set the quality first.

>> The resolution and video quality choices are more limited on the front-facing camera because it's not as sophisticated as the rear camera.

- A picture's resolution describes how many pixels, or dots, are in the image. The more dots, the better the image looks when enlarged.

- An aspect ratio describes the image's overall dimensions horizontally by vertically. A 4:3 ratio is pretty much standard for photographs. The 16:9 ratio is the widescreen format.

- The video quality settings HD and SD refer to high definition and standard definition, respectively. The "p" value represents vertical resolution, with higher values indicating higher quality.

TECHNICAL STUFF

- *Megapixel* is a measurement of the amount of information stored in an image. A megapixel is approximately 1 million pixels, or individual dots that compose an image. It's often abbreviated MP.

Shooting yourself

Why not take advantage of your phone's front-facing camera and take a self-portrait or video? The vernacular term is *selfie*, for self-shot. Get your duck face ready!

Locate the Switch Cameras icon on the Camera app's screen. Figure 13-3 lists a variety of Switch Camera icons I've collected over the years. The one your phone's Camera app uses may look similar.

FIGURE 13-3: Switch Camera icons.

TIP

When you see yourself in the Camera app's viewfinder, you've completed the task successfully. Smile. Click.

Tap the same icon again to switch back to the rear camera. The icon may change its appearance, but you should find it in the same location on the Camera app's screen.

Your Digital Photo Album

The stock Android app for managing photos and videos is called Photos. It lets you peruse the results of your efforts after using the Camera app. It may also bring in photos from your online Google Photos library, which includes images taken by any other Android gizmos you own.

» Beyond viewing photos, you can also perform basic photo-editing chores in the Photos app.

» The Photos app also offers the capability to trim the start or end from a video.

» The traditional Android photo-management/album app is called Gallery. Your phone may sport that app and the Photos app or only one of them. These apps work similarly, though the Photos app offers more features and better integration with online photo sharing.

Viewing your photos and videos

The Photos app organizes your photos and videos in several ways. The Photos screen, shown center in Figure 13-4, lists photos by date. Tap the Photos icon at the bottom of the screen to see photos and videos listed this way. Choose Albums to view any photo albums you've created.

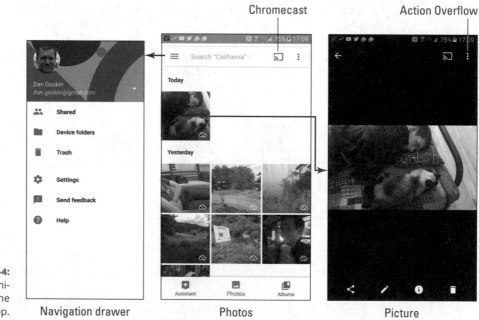

FIGURE 13-4:
Image organization in the Photos app.

Navigation drawer   Photos   Picture

Tap an image to view it full-screen, as shown on the right in Figure 13-4. You can then swipe the screen left or right to browse your images.

Videos stored in an album appear with the Play icon. Tap that icon to play the video. As the video is playing, tap the screen again to view onscreen controls.

>> While you're viewing an image or video full-screen, the navigation icons may disappear. Tap the screen to view them.

>> Tap the Back navigation icon to return to an album after viewing an image or a video.

Starting a slideshow

The Photos app can display a slideshow of your images, but without the darkened room and sheet hanging over the mantle. To view a slideshow, first view an image full-screen. Tap the Action Overflow and choose Slideshow. Images from that particular album or date appear one after the other on the screen.

Tap the Back navigation icon to exit the slideshow.

Slideshows don't have to remain on your phone's diminutive screen: If a nearby HDMI TV or monitor features a Chromecast dongle, tap the Chromecast icon, as shown in the margin. Choose a specific Chromecast device from the list to view the slideshow on the big screen.

To end a Chromecast slideshow, tap the Chromecast icon again and tap the DISCONNECT button. See Chapter 18 for more information on the Chromecast icon.

Finding a picture's location

In addition to snapping a picture, your phone records the specific spot on Planet Earth where the picture was taken. This feature is called Location Tags, GPS-Tag, or Geo-Tag.

To view location-tag information in the Photos app, heed these directions:

1. **View the image.**

2. **Tap the Info icon below the image.**

    On some phones, tap the Action Overflow and choose Details.

The Info card that's displayed shows details about when, how, and where the image was taken, similar to what's shown in Figure 13-5. Map information, if available, appears on the card.

On some phones, you might be able to tap the map and view the location in the Maps app. This feature isn't available on every phone.

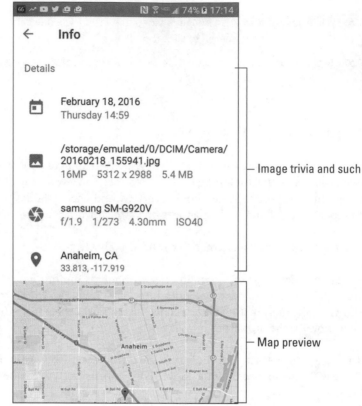

— Image trivia and such

— Map preview

FIGURE 13-5:
Image details,
including
location.

The Location Tags feature isn't without controversy. Some people find it a security risk because the location data is saved with the photo. If you'd rather not have
that information included, heed these directions in the Google Camera app:

1. **Tap the Side Menu icon and choose Settings.**

The goal is to locate the Settings action, which might also be found on the
Action Overflow or appear as a Settings icon on the Camera app's screen.

In some Camera apps, the Settings icon is available without your having to
display the shooting modes.

2. **Enable the Save Location or Location Tags feature.**

Use the master control to enable the feature — or disable it if you don't want
location information saved with your photos.

REMEMBER

The location-tag information is stored in the picture itself. Deactivating this
Location Tags feature doesn't remove that information from photos you've
already taken.

Sharing your images

 Just about every app wants to get in on the sharing bit, especially when it comes to pictures and videos. The key is to view an item in the Photos app and then tap the Share icon, shown in the margin. Choose an app to share the image or video, and that item is instantly sent to that app.

What happens next?

That depends on the app. For Facebook, Twitter, and other social networking apps, the item is attached to a new post. For Gmail, the item becomes a message attachment. Other apps treat images and videos in a similar manner, somehow incorporating the item(s) into whatever wonderful thing that the app does. The key is to look for that Share icon.

Posting a video to YouTube

The best way to share a video is to upload it to YouTube. As a Google account holder, you also have a YouTube account. Why not populate that account with your latest, crazy videos? Who knows what may go viral next!

To upload a video you've recorded, follow these steps:

1. **Ensure that the Wi-Fi connection is activated.**

    The best way to upload a video is to use the Wi-Fi connection, which won't incur data surcharges. In fact, when you use the 4G LTE network for uploading a YouTube video, you see a suitable reminder about the data surcharges.

2. **Open the Photos app.**

3. **View the video you want to upload.**

    You do not need to play the video. Just have it on the screen.

4. **Tap the Share icon.**

    If you don't see the Share icon, tap the screen.

5. **Choose YouTube.**

    You may see a tutorial on trimming the video, which is the next step.

6. **Trim the video, resetting the starting and ending points.**

    This video-editing step is optional. If you opt to trim, adjust the video's starting and ending points left or right. As you drag each marker, the video is scrubbed, allowing you to preview the start and end points.

7. **Type the video's title.**

8. **Set other options.**

   Type a description, set the privacy level, add descriptive tags, and so on.

9. **Tap the Send icon.**

   The video is uploaded to your YouTube account, which may take some time to complete.

The Uploading notification appears while the video is being sent to YouTube. Feel free to do other things with your phone while the video uploads. When the upload has completed, the Uploading notification stops animating and becomes the Uploads Finished notification.

>> When the video is ready for viewing on YouTube, you receive a Gmail message, complete with a link to the new video.

>> To view your video, open the YouTube app. See Chapter 15 for details.

Image Management

The Photos app offers tools for basic image management, editing, and adding some simple effects. I could write an entire book on this app and probably not cover everything you can do. Rather than pester my publisher, I'll constrain myself and cover some of the more basic image-editing options.

Backing up images and videos

The Photos app is preset to automatically back up your phone's images and videos to Google Photos on the Internet, or what's often called "cloud storage." This process takes place automatically as you take photos and record videos.

Visit Google Photos at photos.google.com. You perform image management on the web page, organize albums, and do all sorts of fun stuff. But the real issue is whether you want the images to be backed up.

If you prefer not to share the images automatically, obey these steps to disable the feature:

1. **In the Photos app, tap the Side Menu icon.**

2. **On the navigation drawer, choose Settings.**

3. **Choose Back Up & Sync.**

4. **Slide the master control by Back Up & Sync to the Off position.**

Changing this setting doesn't affect any images already backed up to Google Photos. You can visit Google Photos in a web browser to check on any photos that are already backed up.

Images and videos backed up to Google Photos are private unless you visit the website and opt to make an item public.

Editing an image

To edit an image in the Photos app, you must activate Image Editing mode. This mode also enables some of the app's special effects and image-editing features, but it all starts when you follow these steps:

1. **Display the image you want to edit or otherwise manipulate.**

2. **Tap the Edit icon.**

The Edit icon is shown in the margin. If you don't see it, tap the screen and it shows up.

Editing tools are presented in three categories, shown at the bottom of the screen and illustrated in Figure 13-6: basic settings, image effects, and crop. The basic settings are shown in the figure, which include an auto-adjustment tool, brightness and contrast, color, and so on.

Changes are applied immediately to the image. Tap the Reset icon to undo a change. To cancel everything, tap the Cancel icon. Or, when you're satisfied with your efforts, tap the SAVE button.

Individual effects may have their own Cancel and Done icons. The Done icon is shown in the margin. Even so, you must tap the SAVE button to save the updated image.

Un-editing an image

The changes you make are directly applied to the image; an original copy isn't retained. To remove any previously applied edits, crops, or rotation effects, view the image in the Photos app and follow these steps:

1. **Tap the Edit icon to edit the image.**

2. **Tap the Action Overflow.**

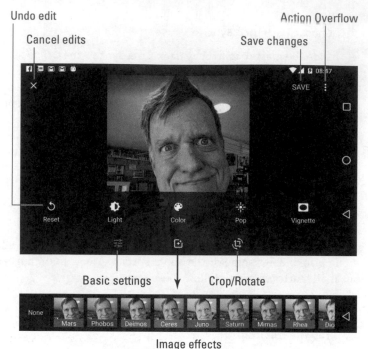

Undo edit

Cancel edits

Action Overflow

Save changes

Basic settings

Crop/Rotate

Reset   Light   Color   Pop   Vignette

None   Mars   Phobos   Deimos   Ceres   Juno   Saturn   Mimas   Rhea   Dio

Image effects

FIGURE 13-6:
Image
editing in the
Photos app.

3. **Choose Undo Edits.**

4. **Tap the SAVE button.**

The original image is restored.

Cropping an image

To *crop* an image is to snip away parts you don't want or need, such as that guy on the far left who photobombed your family picture. To crop an image, obey these steps:

1. **View the image in the Photos app.**

2. **Tap the Edit icon.**

3. **Tap the Crop / Rotate icon.**

    The icon is shown in the margin. The screen changes as illustrated in Figure 13-7. The tools that are presented crop and rotate the image.

Cropping corners   Cropping rectangle

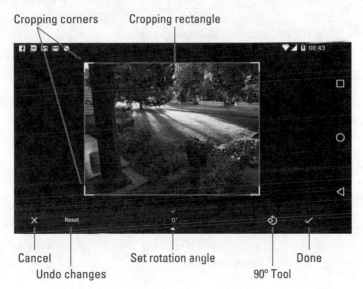

FIGURE 13-7:
Rotating and
cropping an
image.

Cancel   Set rotation angle   Done
  Undo changes   90° Tool

4. **Drag any of the four corners to crop the image.**

   As you drag, portions of the image are removed. You can also drag the image within the cropping rectangle to modify the crop action.

5. **Tap the Done button.**

   The image is cropped. You can continue to edit, or tap the SAVE button to make the changes permanent.

If you're unhappy with the changes after tapping the Done button, tap the Action Overflow and choose Undo Edits.

Rotating pictures

Showing someone else an image on your phone can be frustrating, especially when the image is a vertical picture that refuses to fill the screen when the phone is in a vertical orientation. To fix this issue, rotate the picture in the Photos app. Follow these steps:

1. **Summon the cockeyed image for editing.**

   Refer to the earlier section "Editing an image" for specific directions.

2. **Choose the Crop / Rotate tool.**

3. **Tap the 90° icon to rotate the image in 90-degree increments, or drag the sliders to set a specific angle.**

   Refer to Figure 13-7 for the location of these buttons on the editing screen.

4. **Tap the Done icon to save the changes.**

   You can continue editing, or tap the SAVE button to make the changes permanent.

Rotating an image to a specific angle also crops the image. This step is necessary to keep the image's aspect ratio.

Deleting photos and videos

It's entirely possible, and often desirable, to remove unwanted, embarrassing, or questionably legal images and videos from the Photos app.

 To banish something to bit hell, tap the Delete (Trash) icon on the screen when viewing an image or a video.

The item isn't really deleted. Instead, it's moved to the Trash album. You might even see a MOVE TO TRASH button after you tap the Delete icon.

>> To view the Trash album, tap the Side Menu icon and choose Trash from the navigation drawer.

>> Items held in the Trash album are automatically deleted after 60 days. To hasten the departure, long-press items in the Trash album and then tap the Delete icon atop the screen.

>> You might not be able to delete certain items in the Photos app. These include pictures and videos imported from your online Google Photos library that may be imported into the phone.

Chapter 14

O Sweet Music!

Push-button phones and tone dialing were introduced in the 1970s. People realized that the tones generated by dialing certain numbers sounded like popular tunes. My parents' home number sounded like "Yankee Doodle." On your Android phone, however, listening to music is much more sophisticated than enjoying dialing tones.

In addition to everything else it does, your Android phone can play music. You can listen to songs you buy online, synchronize them from a computer, or find music on the Internet. Some phones can even listen to FM radio. So, wherever you go with your phone, which should be everywhere, you also carry your music library.

The Hits Just Keep On Comin'

Your Android phone is ready to entertain you with music whenever you want to hear it. Simply plug in the headphones, summon the music-playing app, and choose tunes to match your mood. It's truly blissful — well, until someone calls you and the phone ceases being a musical instrument and returns to being the ball-and-chain of the digital era.

To handle music-playing duties, your phone comes with a music-playing app. The stock Android app is Google's own Play Music app.

Browsing the music library

To view your music library, heed these directions:

1. **Start the Play Music app.**

2. **Tap the Side Menu icon to display the navigation drawer.**

   The Side Menu icon is found in the upper left corner of the screen. It's shown in the margin. If you see a left-pointing arrow instead, tap that arrow until the Side Menu icon appears.

3. **Choose Music Library.**

The Play Music app is shown in Figure 14-1 with the Music Library view selected. Your music is organized by categories, which appear as tabs atop the screen. Switch categories by tapping a tab, or swipe the screen left or right to browse your music library.

The categories make your music easier to find, because you don't always remember song, artist, or album names. The Genres category is for those times when you're in the mood for a certain type of music but don't know, or don't mind, who recorded it.

>> Choose the Listen Now item from the navigation drawer to browse songs you frequently listen to or to discover tunes that the Play Music app guesses you'll like. The more you use the app, the more you'll appreciate the results shown in the Listen Now category.

>> Songs and albums feature the Action Overflow icon, similar to the one shown in the margin. Use that icon to view a list of actions associated with the album or artist.

>> The Play Music app shows two types of album artwork. For purchased music, or music recognized by the app, original album artwork appears. Otherwise, the app shows a generic album cover.

>> When the Play Music app doesn't recognize an artist, it uses the title Unknown Artist. This usually happens with music you copy manually to your phone as well as with audio recordings you make.

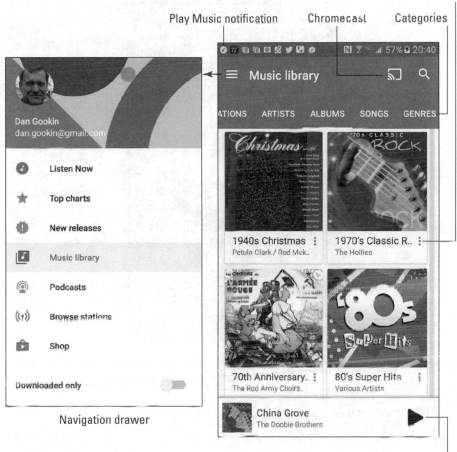

Action Overflow

Play Music notification   Chromecast   Categories

Navigation drawer

Current song

FIGURE 14-1:
The Play Music
app; Albums
category.

Playing a tune

After you've found the proper tune to enhance your mood, play it! Tap on a song
to play that song. Tap on an album to view songs in the album, or tap the album's
large Play button, similar to the one shown in the margin, to listen to the entire
album.

While a song plays, controls appear on a card at the bottom of the screen, shown
in the bottom right in Figure 14-1. Tap that card to view the song full-screen, as
shown in Figure 14-2.

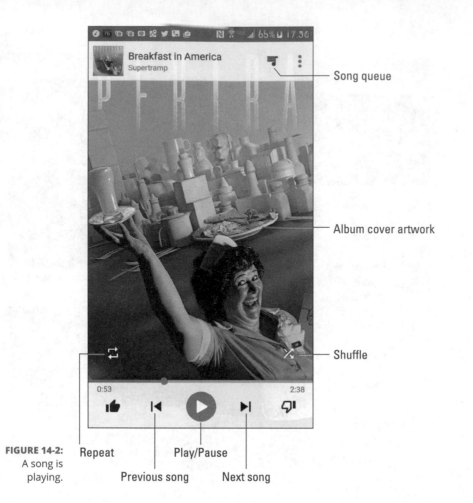

Song queue

Album cover artwork

Shuffle

FIGURE 14-2: Repeat   Play/Pause

A song is

playing.   Previous song   Next song

After the song has finished playing, the next song in the list plays.

The next song doesn't play if you have the Shuffle feature activated. In that case, the Play Music app randomly chooses another song from the same list. Who knows which one is next?

The next song might also not play when you have the Repeat option on: The three Repeat settings are illustrated in Table 14-1, along with the two Shuffle settings. To change settings, tap the Shuffle icon or the Repeat icon to cycle through the settings.

TABLE 14-1

Shuffle and Repeat Icons

Icon	Setting	Effect
X	Shuffle is off.	Songs play one after the other.
X	Shuffle is on.	Songs play in random order.
↻	Repeat is off.	Songs don't repeat.
↻	Repeat current song.	The same song plays over and over.
↻	Repeat all songs.	All songs in the list play over and over.

To stop the song from playing, tap the Play / Pause icon, illustrated in Figure 14-2.

REMEMBER

>> The volume is set by using the Volume key on the side of the phone: Up is louder, down is quieter.

>> Music on your Android phone is streamed from the cloud. That means music plays only when an Internet connection is available.

>> You can download music to play it on your phone without an Internet connection. Directions are offered in Chapter 16.

>> Use the Play Music app's Search icon to help locate tunes in your music library. You can search by artist name, song title, or album. Tap the Search icon (refer to Figure 14-1), and then type all or part of the text you're searching for. Tap the Search key on the onscreen keyboard to start the search. Choose the song you want to hear from the list that's displayed.

>> When a song is playing or paused, its album artwork might appear as the lock screen wallpaper. Don't let the change alarm you.

Controlling the music

The Play Music app keeps you audibly amused until the last song in the list plays, or forever when the Repeat All option is chosen. You're free to do anything else on the phone while music plays: Open another app, read an eBook, or even lock the phone and let the music play.

The only time music is interrupted is for an incoming call. After the call ends, you must restart the song: Visit the Play Music app and tap the Play icon.

 Access the Play Music notification, shown in the margin, to quickly view music controls. Play/Pause and Previous/Next icons are found on the notification drawer, or you can tap the Play Music notification item itself to access the Play Music app.

Queuing up the next song

It's fun to randomly listen to your music library, plucking out tunes like a mad DJ. Oftentimes, however, you don't have patience to wait for the song to finish before choosing the next tune. The solution is to add songs to the queue. Here's how:

1. **Browse your music library for the next song (or album) you want to play.**

2. **Tap the song's Action Overflow.**

3. **Choose Add to Queue.**

   The Play Music app adds the song to the list of tunes to play next.

Songs are added to the queue in the order you tap them. That is, unless you instead choose the Play Next command in Step 3, in which case the tune is inserted next-in-line in the queue.

To review the queue, tap the Song Queue icon, shown in the margin as well as in Figure 14-2. The queue is shown in Figure 14-3.

Songs in the queue play in order, top down. To change the order, drag a song card up or down. To remove a song from the queue, swipe its card left or right.

 If you like your queue, consider making a playlist of those same songs. See the section "Creating a playlist from a song queue," later in this chapter.

Being the life of the party

You need to do four things to make your Android phone the soul of your next shindig or soirée:

>> Connect it to external speakers.

>> Use the Shuffle setting.

>> Set the Repeat option.

>> Provide plenty of drinks and snacks.

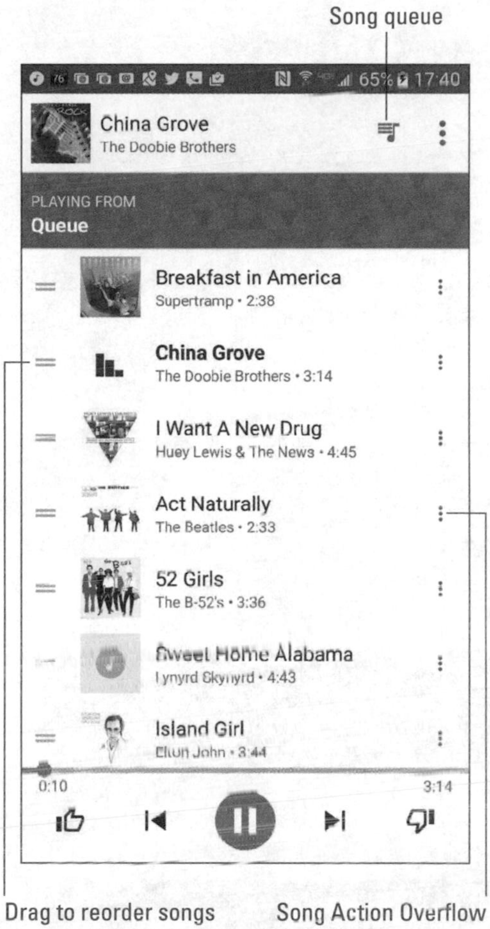

Song queue

PLAYING FROM
Queue

China Grove
The Doobie Brothers

Breakfast in America
Supertramp • 2:38

China Grove
The Doobie Brothers • 3:14

I Want A New Drug
Huey Lewis & The News • 4:45

Act Naturally
The Beatles • 2:33

52 Girls
The B-52's • 3:36

Sweet Home Alabama
Lynyrd Skynyrd • 4:43

Island Girl
Elton John • 3:44

0:10   3:14

FIGURE 14-3:
The song
queue.   Drag to reorder songs   Song Action Overflow

The external speakers can be provided by anything from a custom media dock, a stereo, or any TV connected to a streaming gizmo.

To connect to a stereo, you need an audio cable with a mini-headphone connector for the phone's headphone jack and an audio jack that matches the output device. Look for any store where the employees wear name tags.

See Chapter 18 for information on screencasting. The Chromecast icon, shown in the margin as well as in Figure 14-1, is used to send the Play Music app's output to the other device.

Enjoy your party, and please drink responsibly.

"WHAT'S THIS SONG?"

You might consider getting a handy, music-oriented widget called Sound Search for Google Play. You can obtain this widget from Google Play and then add it to the Home screen, as described in Chapter 16. From a Home screen, you can use the widget to identify music playing within earshot of your phone.

To use the widget, tap it on the Home screen. The widget immediately starts listening to your surroundings, as shown in the middle of the sidebar figure. After a few seconds, the song is recognized and displayed. You can choose to either buy the song at Google Play or tap the Cancel icon and start over.

The Sound Search widget works best (exclusively, I would argue) with recorded music. Try as you might, you cannot sing into the thing and have it recognize a song. Humming doesn't work, either. I've tried playing the guitar and piano and — nope — that didn't work either. But for listening to ambient music, it's a good tool for discovering what you're listening to.

More Music for Your Phone

Your Android phone already has some music preinstalled, thanks to Google and the Play Music service. This random sampling of tunes may not be to your liking, and it's definitely not the final say in how much music you can have on your phone.

To add more music, you have two options:

» Buy lots of music from Google Play, which is what Google wants you to do.

» Borrow music from your computer, which Google also wants you to do, just not as enthusiastically as the first option.

For information on buying music at Google Play, see Chapter 16. To get music from your computer to the Google cloud and make it available to your phone, follow these steps:

1. **On your computer, locate the music you want to upload to your Google Play Music library.**

   You can open a music jukebox program, such as Windows Media Player, or just have a folder window open that lists the songs you want to copy.

2. **Open the computer's web browser.**

3. **Visit** https://music.google.com

   You see a copy of your Play Music library, including your playlists and any recently played songs. You can even listen to your music right there on the computer, but no: You have music to upload.

4. **Click the Side Menu icon on the web page, and then choose the Upload Music option.**

   The Side Menu icon is located in the upper left corner of the web page. You have to scroll down a bit to locate the Upload Music item.

5. **If you haven't yet done so, download the Music Manager.**

   Click the Download Music Manager button. Complete the installation process. You need to perform this task only once.

6. **Drag music into the web browser window, or click the SELECT FROM YOUR COMPUTER button to add tunes.**

   It takes a while for Google to digest the songs, so be patient.

You can repeat these steps to upload tens of thousands of songs. The limit was once 25,000, but I believe Google increased that number recently.

The songs you upload are available to your Android phone, just like any other songs in the music library.

Organize Your Tunes

The Play Music app categorizes your music by album, artist, song, and so forth, but unless you have only one album and enjoy all the songs on it, that configuration probably won't do. To better organize your music, you can create playlists. That way, you can hear the music you want to hear, in the order you want, for whatever mood hits you.

Reviewing your playlists

To view any playlists that you've already created, or that have been preset on the phone, heed these directions:

1. Tap the Side Menu icon in the Play Music app.

2. Choose Music Library.

3. Tap the Playlists tab on the Music Library screen.

Playlists you've created are displayed on the screen, similar to what's shown in Figure 14-4.

To see which songs are in a playlist, tap the playlist's Album icon. To play the songs in the playlist, tap the first song in the list.

TIP

A playlist is a helpful way to organize music when a song's information may not have been completely imported into the phone. For example, if you're like me, you probably have a lot of songs labeled Unknown. A quick way to remedy that situation is to organize those unknown songs on the playlist. The next section describes how it's done.

Creating a playlist

The Play Music app features three "auto" playlists, created automatically by the Play Music app:

Thumbs Up, which lists songs you've liked by tapping the Thumbs Up icon

Last Added, which includes songs recently purchased or added to the phone

Free and Purchased, which includes just about everything

Playlist card

Playlist tab

Playlist Action Overflow

FIGURE 14-4:
Playlists in the
Play Music app.

Current song

Playlist name

To create your own, custom playlist, follow these steps:

TIP

1. **Locate the song you want to add to a playlist.**

Choose your Music Library from the navigation drawer, and then use the tabs
to browse by music category.

You can also add an entire album of songs by choosing an album.

2. **Tap the song's Action Overflow icon.**

3. **Choose Add to Playlist.**

   Ensure that you're viewing a song or an album; otherwise, the Add to Playlist action isn't available.

4. **Choose NEW PLAYLIST.**

   The New Playlist card appears.

5. **Fill in the New Playlist card.**

   Type the playlist's name and a description. You can set the playlist to be public by activating the Public master control. Such playlists can be shared with others on the Internet.

6. **Tap the CREATE PLAYLIST button.**

   The playlist is created and stocked with the song or album you selected in Step 1.

See the next section for details on adding more songs to the playlist.

> » You can have as many playlists as you like on the phone and stick as many songs as you like into them. Adding songs to a playlist doesn't affect the phone's storage capacity.
>
> » To delete a playlist, tap the Action Overflow icon on the playlist's card. (Refer to Figure 14-4.) Choose Delete and tap OK to confirm.

Adding songs to a playlist

To add new songs or albums to an existing playlist, follow these steps:

1. **Tap the Action Overflow icon on the song or album's card.**

2. **Choose Add to Playlist.**

3. **Select the playlist from the list on the Add to Playlist card.**

Repeat these steps to build up existing playlists.

> » To remove a song from a playlist, open the playlist and tap the Action Overflow icon by a song's card. Choose Remove from Playlist.
>
> » You can also swipe a song's card left or right to remove it from the playlist.
>
> » Removing a song from a playlist doesn't delete the song from the music library; see the later section "Deleting music."
>
> » Songs in a playlist can be rearranged: Drag the song's card up or down in the list.

Creating a playlist from a song queue

If you've created a song queue and it's a memorable one, consider saving that queue as a playlist that you can listen to over and over. Obey these directions:

1. **Tap the Song Queue icon to view the song queue.**

   Refer to the earlier section "Queuing up the next song" for details on the song queue.

2. **Tap the Action Overflow icon and choose Save Queue.**

3. **Choose NEW PLAYLIST.**

   Or you can add the songs to an existing playlist, as described in the preceding section.

4. **Fill in the New Playlist card.**

   Refer to the earlier section "Creating a playlist."

5. **Tap the CREATE PLAYLIST button.**

The songs in the current queue now dwell in their own playlist, always accessible.

Deleting music

To remove a song or an album from your Play Music library, tap its Action Overflow icon and choose Delete. Tap the OK button to remove the song. Bye-bye, music.

TIP

I don't recommend removing music. Most music on your phone is actually stored in the cloud, on Google's Play Music service. Therefore, removing the music doesn't affect the phone's storage. So, unless you totally despise the song or artist, removing the music has no effect.

Some music can be stored locally by downloading it to the phone. That way, the music is always available. If the phone's storage capacity is a concern, you can unload the music from the phone. See Chapter 16 for details.

Music from the Stream

Although they're not broadcast radio stations, some sources on the Internet — Internet radio sites — play music. These Internet radio apps are available from Google Play. Some free services that I can recommend are

>> Pandora Radio

>> Spotify

>> TuneIn Radio

Pandora Radio and Spotify let you select music based on your mood and preferences. The more feedback you give the apps, the better the music selections.

The TuneIn Radio app gives you access to hundreds of Internet radio stations broadcasting around the world. They're organized by category, so you can find just about whatever you want. Many of the radio stations are also broadcast radio stations, so odds are good that you can find a local station or two, which you can listen to on your phone.

These apps, as well as other, similar apps, are available for free. Paid versions might also be found at Google Play. The paid versions generally provide unlimited music with no advertising.

>> Google offers an unlimited music listening service. You can sign up by tapping the SUBSCRIBE NOW button, found at the bottom of the Play Music app's navigation drawer. The service is free for 30 days, and then a nominal fee, currently $9.99, is charged monthly.

>> It's best to listen to Internet radio when your phone is connected to the Internet via a Wi-Fi connection. Streaming music doesn't consume a lot of your mobile data plan's monthly allotment, but it will over time.

WARNING

>> Be wary of music subscription services offered through the phone's manufacturer or cellular provider. I've subscribed to such services only to find them terminated for various reasons. To avoid that disappointment, stick with the services described in this section until you feel comfortable enough to buy into another service.

>> Some phones may feature FM radio apps. These apps magically pull radio signals from the air and put them into your ear. The phone's radio hardware requires that a headset be plugged in for the app to work.

TECHNICAL STUFF

>> Internet music of the type delivered by the apps mentioned in this section is referred to by the nerds as *streaming* music. That's because the music arrives on your phone as a continuous download from the source. The music is played as it comes in and isn't stored long-term.

Chapter 15

Apps Various and Sundry

When cell phones started to become sophisticated, manufacturers began installing basic programs, or apps. Those early apps were pitiful. In fact, I remember how my first cell phone came with a smattering of boring, silly little apps. Still, it's kind of traditional for a phone to come with an alarm clock, a calculator, a calendar, and perhaps a few games. Your Android phone is no different.

The Alarm Clock

Your Android phone keeps constant and accurate track of the time, which is displayed at the top of the Home screen and also the lock screen. That's handy for telling the time, but not for any time tricks. Therefore, the phone ships with a Clock app.

The Clock app is your phone's chronometric app, featuring a timer, a stopwatch, an alarm, and world clock functions. Of these activities, setting an alarm is pretty useful: In that mode, your phone becomes your nightstand companion — and, potentially, your early morning nemesis.

To set an alarm in the Clock app, follow these generic steps:

1. **Tap the Alarm icon or tab atop the Clock app's screen.**

   The four tabs in the stock Android Clock app are Alarms, World Clock, Timer, and Stopwatch, although not specifically in that order.

2. **Tap the Add icon or ADD button.**

   A card appears, which you use to set the alarm time, days, name, and so on.

3. **Fill in details about the alarm.**

   Set the alarm's time. Determine whether it repeats daily or only on certain days. Choose a ringtone. Weigh over any other settings shown on the card. The alarm name appears when the alarm triggers.

4. **Set the alarm.**

   Alarms must be set to activate.

REMEMBER

When the alarm triggers, its name appears on the lock screen. Tap the Dismiss icon and actually get out of bed. Or, you can tap the Snooze icon to be annoyed again after a few minutes.

>> On some phones, you may find the Alarm app instead of the Clock app. As its name suggests, the Alarm app only sets alarms.

>> When an alarm is set, the Alarm status icon appears atop the screen, similar to what appears in the margin. The icon is your clue that an alarm is set and ready to trigger.

>> Un-setting an alarm doesn't delete the alarm. To remove an alarm, tap the alarm and then tap the Delete (Trash) icon. In some incarnations of the Clock app, long-press an alarm and choose the Delete action.

>> The alarm doesn't work when you turn off the phone. However, the alarm does trigger when the phone is locked.

>> For a larger time display, you can add the Clock widget to the Home screen. Refer to Chapter 19 for more information about widgets on the Home screen.

>> So tell me: Do alarms go off, or do they go on?

The Calculator

The Calculator is perhaps the oldest of all traditional cell phone apps. It's probably also the least confusing and frustrating app to use.

The stock Android calculator app appears in Figure 15-1. The version you see on your phone may look different, although the basic operation remains the same. Also, consider changing the phone's orientation to see more or fewer buttons; the image in Figure 15-1 uses horizontal orientation, which shows the calculator's more terrifying buttons.

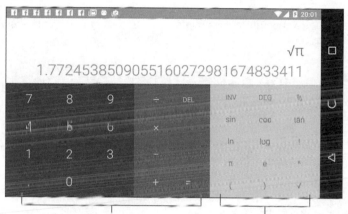

FIGURE 15-1:
The Calculator.

Typical calculator buttons   Scary calculator buttons

Tap the various buttons on the Calculator app screen to input your equation. Some buttons, such as √ (square root) must come before you type a value, as illustrated in the figure.

TIP

» Use the parentheses to determine which part of a long equation gets calculated first.

» Long-press the calculator's text (or results) to cut or copy the results. This trick may not work in every Calculator app.

» To clear the Calculator app display, long-press the DEL, CLR, or C button.

The Calendar

Once upon a time, people toted around a bulky notebook thing called a *datebook*. It assisted primitive humans with keeping a schedule, reviewing appointments, and knowing where they need to be and when. Such archaic technology is no longer necessary, because your Android phone comes with a Calendar app.

TIP

>> The Calendar app works with your Google account to keep track of your schedule and appointments. You can visit Google Calendar on the web at

    http://calendar.google.com

>> Before you throw away your datebook, copy into the Calendar app some future appointments and info, such as birthdays and anniversaries.

Browsing dates and appointments

To check your schedule and browse events, open the Calendar app. You see upcoming dates shown in one of several views; Figure 15-2 shows the Calendar app's Month, Week, and Day views. Not shown are 3 Day view or Schedule view, which simply lists upcoming events.

FIGURE 15-2:
The Calendar app.

To change views, tap the Side Menu icon and choose a view type from the navigation drawer.

Swipe the screen left or right to browse events. If you need to return to today's date, tap the Go to Today icon, illustrated in Figure 15-2.

>> Not every version of the Calendar app displays your schedule exactly as illustrated in Figure 15-2. Some versions may lack Week view or 3 Day view, for example. Others may sport an action bar from which the current calendar view is chosen.

>> When using the 3 Day, Day, Week, or Schedule views, tap the month name to see a quick overview of the current month's appointments.

>> I check Week view at the start of the week to remind me of what's coming up.

>> The current day is highlighted in Month and Week views. A horizontal bar marks the current time. (Refer to Day view in Figure 15-2.)

>> Different colors flag your events. The colors represent a calendar category to which the events are assigned. See the later section "Creating an event," for information on calendars.

Reviewing appointments

To view more detail about an event, tap it. When you're using Month view, first tap the event's date and then the event. The event's details appear onscreen, similar to the event shown in Figure 15-3.

Which details you see depends on how much information was recorded when the event was created. Some events have only a minimum of information; others may have details, such as a location for the event. When the event's location is listed, you can tap that location on the card, and the Maps app pops up to show you where the event is being held.

Tap the Back navigation icon to dismiss the event's details.

>> Birthdays and a few other events on the calendar may be pulled from the phone's address book or from social networking apps. That coordination explains why some events are listed twice; they're pulled in from two sources.

>> The best way to review upcoming appointments is to choose Schedule view.

>> A Calendar widget also provides a great way to see upcoming events from directly on the Home screen. See Chapter 19 for information on widgets.

>> Google Now also lists any immediate appointments or events. See the later section "Google Now."

Edit event

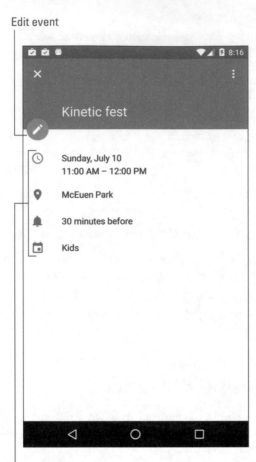

FIGURE 15-3:
Event details.

Event details

Creating an event

The key to making the calendar work is to add events: appointments, things to do, meetings, or full-day events such as birthdays or root canal work. To create an event, follow these steps in the Calendar app:

1. **Select the day for the event.**

Or, if you like, you can switch to Day view, where you can tap the starting time for the new event.

2. **Tap the New Event icon.**

3. **If prompted, choose Event as the item you want to add.**

Upon success, you see the New Event card. Your job is to fill in the blanks to create the event.

4. **Set information for the event.**

The more information you supply, the more detailed the event, and the more you can do with it on your phone. Here are some of the many items you can set when creating an event:

- *Title:* The name of the event, person you're going to meet, or place you're headed.

- *Calendar Category:* Choose a specific calendar to help organize and color-code your events. If you haven't configured calendar categories, this choice doesn't appear.

- *Time/Duration:* If you followed Step 1 in this section, you don't have to set a starting time. Otherwise, specify the event's date, starting time, and ending time. You can also set an all-day event such as a birthday or the while installer's visit, although they claimed he'd show up between 8:00 and 11.00.

- *Location:* Type the location just as though you're searching for a location in the Maps app.

- *Repeat:* Configure the event to happen regularly on a monthly or weekly schedule. If you don't see the item, tap the More Options button.

- *Notification/Reminder:* Set an email, text message, or Calendar notification to signal an upcoming event.

5. **Tap the Save or Done button to create the event.**

The new event appears on the calendar, reminding you that you need to do something on such-and-such a day.

When an event's date and time arrives, an event reminder notification appears, similar to the one shown in the margin. You might also receive a Gmail notification or text message, depending on how you chose to be reminded when the event was created.

» To change, update, or edit an event, tap the event to bring up its card, as shown earlier, in Figure 15-3. Tap the Edit icon to make modifications.

TIP

>> For events that repeat twice a week or twice a month, create two repeating events. For example, when you have meetings on the first and third Mondays, create one event for the first Monday and another for the third. Then have each event repeat monthly.

>> To remove an event, first tap the Edit icon on the event's card to edit the event. Tap the Delete (Trash) icon or DELETE button. Tap OK to confirm. For a repeating event, choose whether to delete only the current event or all future events.

>> Setting an event's time zone is necessary only when the event takes place in another time zone or spans time zones, such as an airline flight. In that case, the Calendar app automatically adjusts the starting and stopping times for events, depending on where you are.

WARNING

>> If you forget to set the time zone and you end up hopping around the world, your events are set according to the time zone in which they were created, not the local time.

>> Avoid using the Phone and Device categories for your events. Events in those categories appear on your phone but aren't shared with your Google account.

TECHNICAL STUFF

>> Calendar categories are handy because they let you organize and color-code your events. They can be confusing because Google calls them "calendars." I think of them more as categories: I have different calendars (categories) for my personal and work schedules, government duties, clubs, and so on.

The eBook Reader

To sate your electronic book-reading desires, your Android phone comes with Google's eBook reader app, Play Books. It offers you access to your eBook library, plus the multitudinous tomes available from Google Play.

Open the Play Books app to behold any eBooks you've recently read, which are displayed on the Read Now screen. To browse your entire eBook library, tap the Side Menu icon and choose My Library from the navigation drawer.

The library lists any titles you've obtained for your Google Books account, similar to what's shown in Figure 15-4.

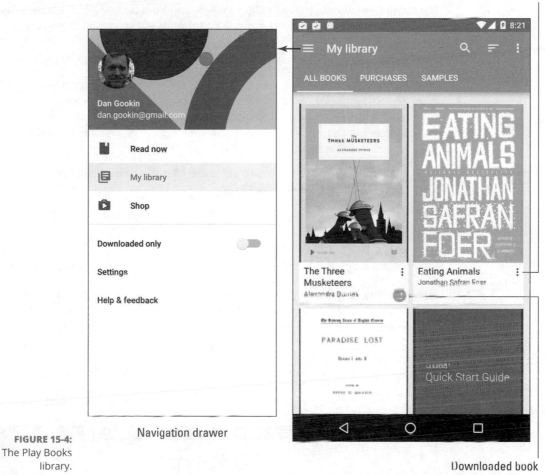

Book's Action Overflow

FIGURE 15-4:
The Play Books
library.

Navigation drawer

Downloaded book

Swipe the screen to scroll through the library.

Tap a book's cover (or card) to open it. If you've previously opened the book, you see the page you last read. Otherwise, you see the book's first page.

Figure 15-5 illustrates the basic book-reading operation in the Play Books app. You swipe the screen right-to-left to turn pages. You can also tap either the far left or right side of the screen to turn pages.

The Play Books app works in both vertical and horizontal orientations. You can lock the screen orientation for the app to either direction: Choose the Settings item from the navigation drawer (refer to Figure 15-4), and then choose Auto-Rotate Screen. Select a Lock item to prevent the app from rotating the screen while you read.

>> If you don't see a particular book in your library, tap the Action Overflow and choose Refresh.

>> Books in your Play Books library are stored on the Internet and available to read only when an Internet connection is active. It's possible to keep a book on your phone by downloading it to the device. Refer to Chapter 16 for details on downloading books.

>> To remove a book from the library, tap the Action Overflow on the book's cover and choose Delete from Library.

>> If the onscreen controls (refer to Figure 15-5) disappear, tap the screen to see them again.

>> Tap the Aa icon to view options for adjusting the text on the screen and the brightness level.

>> Unlike dead-tree books, eBooks lack an index. That's because text on digital pages can change based on the book's presentation. Therefore, use the Search icon (refer to Figure 15-5) to find text.

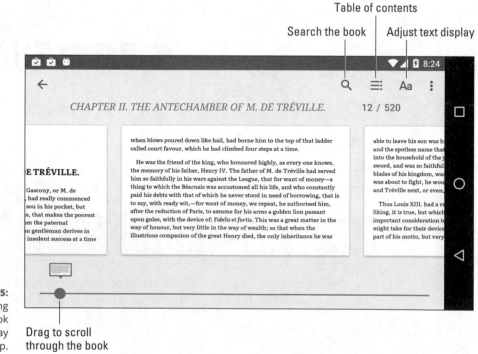

Table of contents

Search the book   Adjust text display

FIGURE 15-5: Reading an eBook in the Play Books app.

Drag to scroll through the book

» Copies of all your Google Books are available on all your Android devices and on the http://books.google.com website.

» Refer to Chapter 16 for information on obtaining books from Google Play as well as how to download books to your phone.

» If you have a Kindle device, you can obtain the Amazon Kindle app for your phone. Use the app to access books you've purchased for the Kindle, or as a supplement to Google Books.

The Game Machine

For all its seriousness and technology, one of the best uses of a smartphone is to play games. I'm not talking about silly arcade games (though I admit that they're fun). No, I'm talking about some serious portable gaming.

To whet your appetite, your Android phone may have come with a small taste of what the device can do in regard to gaming; look for preinstalled game apps in the Apps drawer. If you don't find any, choose from among the hordes available from Google Play, covered in Chapter 16.

Game apps use phone features such as the touchscreen or the accelerometer to control the action. It takes some getting used to, especially if you regularly play using a game console or PC, but it can be fun — and addicting.

Google Now

Don't worry about the phone controlling too much of your life: It harbors no insidious intelligence, and the Robot Uprising is still years away. Until then, you can use your phone's listening capabilities to enjoy the feature called Google Now. It's not quite like having your own personal Jeeves, but it's on its way.

Finding the Google Now feature can be fun! On some phones, you drag your finger on the touchscreen from bottom to top to summon it. Other phones may feature the Google Now screen as the far left Home screen page. Or you can tap the Google Search widget on the Home screen to summon Google Now.

Figure 15-6 illustrates a typical Google Now screen. Cards appear below the search text box. The variety and number of cards depend on how often you use Google Now. The more it learns about you, the more cards appear.

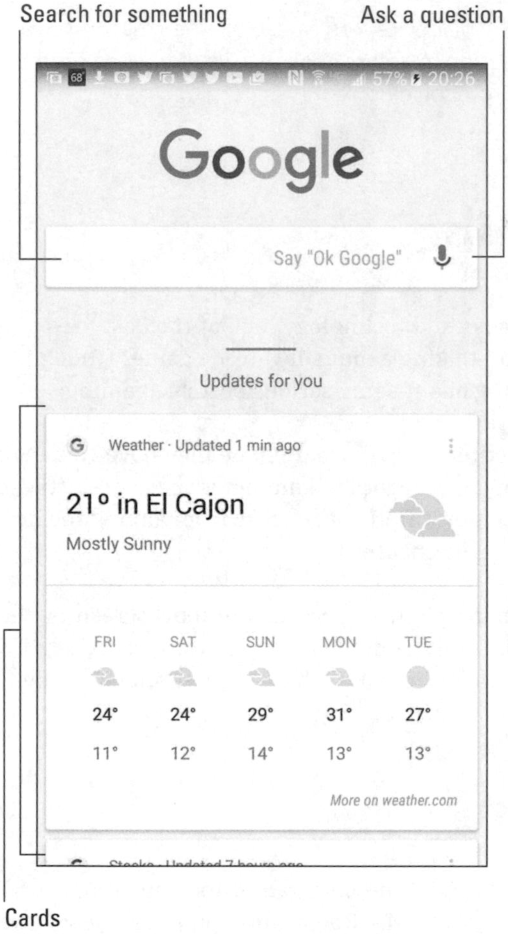

Search for something   Ask a question

Google

Say "Ok Google"   🎤

Updates for you

G   Weather · Updated 1 min ago   ⋮

21° in El Cajon

Mostly Sunny

FRI	SAT	SUN	MON	TUE
24°	24°	29°	31°	27°
11°	12°	14°	13°	13°

More on weather.com

Stocks · Updated 7 hours ago

FIGURE 15-6:
Google Now
is ready for
business.
Or play.   Cards

You can use Google Now to search the Internet, just as you would use Google's main web page. More interesting than that, you can ask Google Now questions; see the nearby sidebar, "Barking orders to Google Now."

Google Now might also dwell in the Apps drawer. It might be named Google Now or just Google. On some phones, you may have to tap the Get Google Now link to obtain the app.

TIP

BARKING ORDERS TO GOOGLE NOW

One way to have a lot of fun is to use the Google Now app verbally. Just say "Okay, Google." Say it out loud. Anytime you see the Google Now app, it's listening to you. Or when the app is being stubborn, tap the Microphone icon.

You can speak simple search terms, such as "Find pictures of Megan Fox." Or you can give more complex orders, among them:

- Will it rain tomorrow?
- What time is it in Frankfurt, Germany?
- How many euros equal $25?
- What is 103 divided by 6?
- How can I get to Disneyland?
- Where is the nearest Sri Lankan restaurant?
- What's the score of the Lakers–Celtics game?
- What is the answer to life, the universe, and everything?

When asked such questions, Google Now responds with a card as well as a verbal reply. When a verbal reply isn't available, Google search results are displayed.

The Video Player

It's not possible to watch broadcast TV live on your phone, but a few apps come close. The YouTube app is handy for watching random, meaningless drivel, which I suppose makes it a lot like broadcast TV. Google offers the Play Movies & TV app, which lets you buy and rent real movies and TV shows from Google Play. And when you tire of those apps, you can use the Camera app with the front-facing camera to pretend that you're the star of your own reality TV show.

Watching YouTube

YouTube is the Internet phenomenon that proves that real life is indeed too boring and random for television. Or is that the other way around? Regardless, you can view the latest videos on YouTube — or contribute your own — by using the YouTube app on your Android phone.

Tap the Search icon to hunt down YouTube videos. Type the video name, a topic, or any search terms to locate videos. Zillions of videos are available.

The YouTube app displays suggestions for any channels to which you're subscribed, which allows you to follow favorite topics or YouTube content providers.

REMEMBER

>> Use the YouTube app to view YouTube videos, rather than use the phone's web browser app to visit the YouTube website.

>> Orient the phone horizontally to view the video in a larger size.

>> Because you have a Google account, you also have a YouTube account. I recommend that you log in to your YouTube account when using the YouTube app: Tap the Action Overflow and choose Sign In. You see your account information, your videos, and any video subscriptions.

>> Not all YouTube videos are available for viewing on mobile devices.

Buying and renting movies

You can use the Play Movies & TV app to watch videos you've rented or purchased from Google Play. Open the app and choose the video from the main screen. Items you've purchased show up in the app's library.

Use the Play Store app to rent and purchase videos. Check there often for freebies and discounts. More details for renting and purchasing movies and shows is found in Chapter 16.

TIP

>> Movies and shows rented from Google Play are available for viewing for up to 30 days after you pay the rental fee. After you start the movie, you can pause and watch it again and again during a 24-hour period.

>> Not every film or TV show is available for purchase. Some are rentals only.

>> Also see Chapter 18 for information on screencasting video from your phone to a large-screen monitor or HDTV.

Chapter 16

Shop at Google Play

Your Android phone may have come packed with diverse and interesting apps, or it may have only a paltry selection of tepid sample apps. Regardless of the assortment, you're in no way limited to the original collection. That's because hundreds of thousands of different apps are available for your phone: productivity apps, references, tools, games, and more. They're all found at the central shopping place for all Android phones: Google Play.

Welcome to the Store

Google Play, sometimes referred to as the Play Store, is one of the things that makes owning an Android phone rewarding. You can visit Google Play to not only obtain more apps for your phone but also go shopping for music, books, and videos (films and TV shows). Someday, Google might even sell robots. Until then, the selection seems adequate and satisfying.

>> Officially, the store is called Google Play. It was once known as the Play Store and the app is named Play Store. In the future, the app name may change to Google Play.

»   Google Play was once known as Android Market, and you may still see it referred to as the Market.

»   You obtain apps and media from Google Play over an Internet connection. Therefore:

»   I highly recommend that you connect your phone to a Wi-Fi network if you plan to obtain apps, books, or movies at Google Play. Wi-Fi not only gives you speed but also helps you avoid data surcharges. See Chapter 17 for details on connecting your phone to a Wi-Fi network.

»   The Play Store app is frequently updated, so its look may change from what you see in this chapter. Refer to my website for updated info and tips: http://wambooli.com/help/android

Browsing Google Play

To access Google Play, open the Play Store app, found in the Apps drawer. You may also find a launcher on the Home screen.

After opening the Play Store app, you see the main screen, similar to the one shown in Figure 16-1. The store is divided into two parts: one titled Apps & Games and another titled Entertainment.

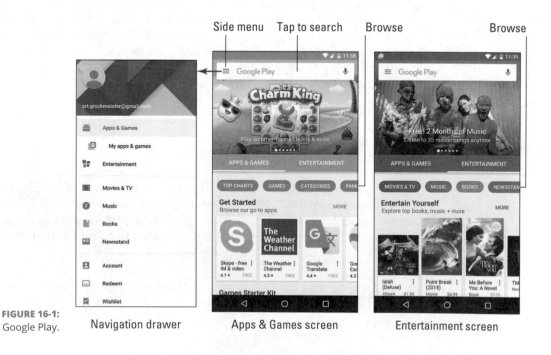

FIGURE 16-1:
Google Play.

Navigation drawer   Apps & Games screen   Entertainment screen

If you don't see the main screen, similar to what's shown in Figure 16-1, tap the Side Menu icon (shown in the margin) to display the navigation drawer. Choose Apps & Games to view the apps portion of the store, or choose Entertainment to view media items.

To browse each top-level category, choose a category. For Apps & Games, you can choose Top Charts, Games, and so on, as illustrated in Figure 16-1. For Entertainment, choose Movies & TV, Music, Books, and so on.

After you browse to a specific item, further categories help you browse. These categories include top sellers, new items, free items, and so on. Eventually you see a list of suggestions, similar to what's shown on the left in Figure 16-2. Swipe the suggestions up and down to peruse the lot.

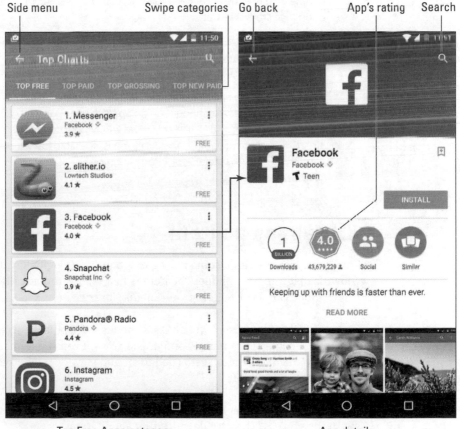

FIGURE 16-2: Hunting down an app.

Tap an item to see a more detailed description, screen shots, a video preview, comments, and links to similar items, as shown on the right in Figure 16-2.

When you have an idea of what you want, such as an album name or an app's function, use the Search command instead: Tap the search box at the top of the screen (refer to Figure 16-1) or use the Search icon, illustrated in Figure 16-2. As you type search text, a list of matching suggestions appears on the screen.

>> The first time you open the Play Store app, or after the app is updated, you have to accept the terms of service. To do so, tap the ACCEPT button.

>> You can be assured that all apps in Google Play can be used with your phone. There's no way to obtain something that's incompatible.

>> Pay attention to an app's ratings. (Refer to Figure 16-2.) Ratings are added by people who use the apps — people like you and me. Having a higher rating out of five stars is better.

>> Another good indicator of an app's success is how many times it's been downloaded. Some apps have been downloaded tens of millions of times. That's a good sign.

TECHNICAL
STUFF

>> In Figure 16-2, the app's description (on the right) shows an INSTALL button. Other buttons that may appear on an app's description screen include OPEN, UPDATE, REFUND, and UNINSTALL. The OPEN button opens an app that's already installed on your phone. A refund is available just after you purchase something. See Chapter 19 for information on using the UPDATE and UNINSTALL buttons.

Obtaining an item

After you locate an app, some music, or a book on Google Play, the next step is to download it into your phone. The item is installed automatically, expanding what your phone can do or building your media library.

Good news: Most apps are free. Classic books are available at no cost. And occasionally, Google offers movies and music *gratis*. Even the items you pay for don't cost that much. In fact, it seems odd to sit and stew over whether paying 99 cents for a game is "worth it."

TIP

I recommend that you download a free app or book first, to familiarize yourself with the process. Then try downloading a paid item.

Free or not, the process of obtaining something from Google Play works pretty much the same. Follow these steps:

1. **If possible, activate the Wi-Fi connection, to avoid incurring data overages.**

   The phone uses Wi-Fi when it's available, preferring it over the mobile data network. See Chapter 17 for information on connecting your phone to a Wi-Fi network.

2. **Open the Play Store app.**

3. **Find the item you want and open its description.**

   All items in the Play Store app feature a description screen. It looks similar to the app description screen shown on the right in Figure 16-2.

4. **Tap the button to obtain the item.**

   A free app features the INSTALL button. A free book features the ADD TO LIBRARY button. For free movies, TV shows, or music, look for the FREE button. You might also see the FREE TRIAL button for some items. In that case, tap the button to view or listen to a free sample of the media.

   Paid items feature a button that shows the price. For movies and TV shows, you may see a RENT or BUY button. See the later section "Renting or purchasing videos."

5. **If prompted, tap the ACCEPT button.**

   The ACCEPT button appears on an access card. It describes which phone features the app uses. The list isn't a warning, and it doesn't mean anything bad. Still, see the later sidebar "Avoiding Android viruses."

6. **For a paid item, tap the BUY button.**

   If you're purchasing an item, you must confirm the payment method. See the next section for details.

7. **Wait for the item to download.**

   Media items are instantly available. Apps are downloaded to your phone, which may take some time. You see the Downloading notification appear atop the screen, similar to the one shown in the margin.

   You're free to do other things on your phone while an app is downloaded and installed.

8. **Tap the OPEN, PLAY, LISTEN, READ, or similar button to run the app, watch a video, listen to music, or read a book, respectively.**

Music, eBooks, movies, and videos are available for instant enjoyment, but not necessarily downloaded to your phone. See the later section "Keeping stuff on the device."

» If you haven't yet supplied payment information for Google Play, the Play Store app may prompt you to enter credit card information, even when you're obtaining a free item. If you don't want to supply payment information, tap the SKIP button.

» To quickly access something you've just acquired from Google Play, look for the Successfully Installed notification, shown in the margin. The notification features the item's name with the text *Successfully Installed* beneath it.

» Some apps don't use any phone features, so an acceptance card doesn't appear. (Refer to Step 5 in this section.)

» Many apps prompt for permissions when they're first opened. Tap ALLOW or DENY. Generally speaking, it's okay to tap ALLOW, but also refer to the later sidebar "Avoiding Android viruses."

» A few apps require that you accept a license agreement. If so, tap the I AGREE button when you first start the app. Additional app setup may involve setting your location, signing in to an account, or creating a profile, for example.

» Media you've obtained from Google Play is accessed from a specific app: Play Music for music, Play Books for books, and Play Movies & TV for video. Other chapters in this part of the book offer details.

» When you buy something from Google Play, you receive a Gmail message confirming your purchase, paid or free. The message contains a link you can select to review the app refund policy, in case you change your mind about the purchase.

» Be quick on that refund: For a purchased app, you have only two hours to get your money back. You know when the time limit is up because the REFUND button on the app's description screen changes to UNINSTALL.

» Google Play doesn't currently offer refunds on purchased media, which includes music, books, and movies.

TIP

» Keep an eye out for special offers from Google Play. These offer a great way to pick up some free songs, movies, and books.

WARNING

AVOIDING ANDROID VIRUSES

How can you tell which apps are legitimate and which might be viruses or evil software that does odd things to your Android phone? Well, you can't. In fact, most people can't, because evil apps don't advertise themselves as such.

The key to knowing whether an app is malicious is to look at its access card: Open the app's description in the Play Store app, and see which permissions it's granted. For example, if a simple grocery-list app uses the phone's microphone and the app doesn't need to use the microphone, it's suspect.

In the history of the Android operating system, only a handful of malicious apps have been distributed, and most of them were found on devices used in Asia. Google routinely removes malicious apps from its inventory. Google is also capable of remotely wiping such apps from all Android devices. So you're pretty safe.

Generally speaking, avoid "hacker" apps, porn apps, and apps that use social engineering to make you do things on your phone that you wouldn't otherwise do, such as visit an unknown website to see racy pictures of politicians or celebrities.

Purchasing something from Google Play

To purchase an app or media on Google Play, you tap the BUY button. A card appears, listing your preferred payment method, such as the example shown in Figure 16-3.

In the figure, the movie *Agora* is listed for $13.77, which is the HD purchase price. The chosen payment method is a Visa card ending in 6998. To use that payment method, follow these steps:

1. **Tap the BUY button.**

For security, you're prompted to type your Google password.

2. **Type your Google password.**

WARNING

I strongly recommend that you *do not* choose the option Never Ask Me Again. You want to be prompted every time for your password.

3. **Tap the Confirm button.**

Current payment method

Chevron

Agora $13.77
Visa-6998 ⌄

Google Play   [BUY]

Agora $13.77
Visa-6998 ⌃

Payment methods

Redeem

art.grockmeister@gmail.com
Includes Tax of $0.78.

Google Play   [BUY]

Cash in a gift card

FIGURE 16-3:
The Buy card.   Choose another form of payment

4. **Type the credit card's security code.**

    This is the CVC code, found on the back of the credit card.

5. **Tap the Verify button.**

    The app is downloaded or the media made available to your phone.

To select another payment method, follow these steps when the Buy card is presented, and tap the chevron by the current payment method, as shown in Figure 16-3. The Buy card expands, illustrated at the bottom of Figure 16-3. Choose Payment Methods and select another credit or debit card or use your Google Play balance. After another payment method is selected, continue with Step 1 in this section.

If you've not yet set up a payment method, the chevron appears by the item's price, not below the item purchased, as shown in Figure 16-3. Tap that chevron, and then choose a payment method. You can add a credit or debit card, bill via your cellular provider, use PayPal, or redeem a Google Play gift card.

NEVER BUY ANYTHING TWICE

Any items you've already purchased from Google Play are available on your Android phone. These include apps, books, music, videos, and other media. So if you have an Android device or are upgrading from an older Android phone, all your paid apps, music, books, and other items are available to your current phone.

To review any already purchased apps, display the Play Store app's navigation drawer. (Refer to the left side of Figure 16-1.) Choose Apps & Games, and then display the navigation drawer again and choose My Apps & Games. Tap the All tab on the My Apps & Games screen to see all apps you've ever obtained from Google Play, including apps you've previously paid for. Those apps are flagged with the text *Purchased*. Choose that item from the list to reinstall the paid app.

» After you purchase the item, it's made available to your phone. Apps are downloaded and installed; music, eBooks, movies, and videos are available but not necessarily downloaded. See the later section "Keeping stuff on the device."

» The credit or debit cards listed in Google Play are those you've used before. Don't worry: Your information is safe.

Renting or purchasing videos

When it comes to movies and TV shows available from Google Play, you have two options: rent or purchase.

When you choose to rent a video, the rental is available to view for the next 30 days. After you start watching, however, you have only 24 hours to finish; during that time, you can watch the video over and over.

Purchasing a video is more expensive than renting it, but you can view the movie or TV show at any time, on any Android device. You can also download the movie so that you can watch it when an Internet connection isn't available.

When you opt to purchase a video, you may be prompted to select the SD or HD version. The SD version is cheaper and occupies less storage space (if you choose to download the movie). The HD version is more expensive, but it plays at high definition only on certain output devices. Obviously, when watching on your phone only, the SD option is preferred.

> >> Refer to Chapter 15 for information on watching movies and TV shows.

> >> Also see Chapter 18 for information on screencasting videos from your phone to a large-screen device, such as an HDTV.

Google Play Tricks

Like most people, you probably don't want to become a Google Play expert. You just want to get the app you want or music you desire and get on with your life. Yet more exists to the Play Store app than simply obtaining new stuff.

Using the wish list

 While you dither over getting a purchase at Google Play, consider adding the item to your wish list: Tap the Wish List icon when viewing the app. The Wish List icon is shown in the margin, although its color changes depending on which category you're viewing in the Play Store app.

To review your wish list, tap the Side Menu icon in the Play Store app. (Refer to Figure 16-1.) Choose the My Wishlist item from the navigation drawer. You see all the items you've flagged. When you're ready to buy, choose one and buy it!

Sharing an item from Google Play

Sometimes you love your Play Store purchase so much that you just can't contain your glee. When that happens, consider sharing the item. Obey these steps:

1. **Open the Play Store app.**

2. **Browse or search for the app, music, book, or other item you want to share.**

3. **When you find the item, tap it to view its description screen.**

4. **Tap the Share icon.**

   You may have to swipe down the screen to locate the Share icon, shown in the margin. After tapping the Share icon, you see a menu listing various apps.

5. **Choose an app.**

   For example, choose Gmail to send a Google Play link in an email message.

6. **Use the chosen app to share the link.**

   What happens next depends on which sharing method you've chosen. For example, if you chose Gmail, compose the message.

The result of following these steps is that your friend receives a link. That person can tap the link on his mobile Android device and be whisked instantly to the Play Store app, where the item can be obtained.

Keeping stuff on the device

Books, music, and video that you obtain from Google Play are not copied to your phone. Instead, they're stored on the Internet. When you access the media, it's streamed into your device as needed. This setup works well, and it keeps your phone from running out of storage space, but it works only when an Internet connection is available.

When you plan on being away from an Internet connection, such as when you are flying cross-country and are too stingy to pay for inflight Wi-Fi, you can download music, eBook, and movie purchases and save them on your phone.

To see which media is on your phone and which isn't, open the Play Books, Play Music, or Play Movies & TV app. Follow these steps, which work identically in each app:

1. **Tap the Side Menu icon.**

2. **In the navigation drawer, choose the Downloaded Only item.**

   The master control by the item is either on or off, depending on the item's setting. If this step just now turned the item off, tap it again to turn it on.

3. **Choose the Library item from the navigation drawer.**

   You see only those items stored directly on your phone. The rest of your library, you can assume, is held on the Internet.

To see your entire library again, repeat these steps, and in Step 2, ensure that the master control is in the Off position.

 Items downloaded to your phone feature the On Device icon, similar to the one shown in the margin. The icon's color differs between music, eBooks, and movies.

To keep an item on your phone, look for the Download icon, similar to what's shown in the margin. Tap that icon, and the item is fetched from the Internet and stored on your phone.

 Keeping movies and lots of music on your Android phone consumes a lot of storage space. That's okay for short trips and such, but for the long term, consider purging some of your downloaded media: Tap the On Device icon. Tap the Remove button to confirm.

WARNING

Don't worry about removing downloaded media. You can always access items in the phone's media library when an Internet connection is active. And you can download items over and over without having to pay again.

Buying something remotely

Google Play also features a website, which can be found at `play.google.com/store`. This online store features the same apps, videos, music, books, and so on that can be found in the Play Store app on your phone. It also includes devices, which are various hardware gizmos — including phones and tablets — you can purchase.

A nifty trick you can pull on the Google Play website is to remotely install apps on your phone or any Android device: Visit the website and click the Sign In button if you haven't yet signed in. Use your Google account, the same one you use on your Android phone.

As you browse items on the Google Play website, you can obtain apps or purchase media just as you would on your phone. An extra step has you choose on which of your Android devices you want the item installed. For example, you can choose an app and then select your Android phone from the Choose a Device menu. Click the Install button, and the device is remotely installed on your phone.

Don't worry! No one else can use this feature to remotely install items on your phone. Only when you use your Google (or Gmail) account to sign in is this service available.

4

Nuts and Bolts

IN THIS PART . . .

Seek out and find Wi-Fi networks, connect, and use the Internet.

Exchange files between your phone and other devices.

Work with apps and widgets on the Home screen.

Customize and configure your phone.

Keep your phone private and secure.

Take your phone abroad.

Maintain and troubleshoot your phone.

Chapter 17

No Wires, Ever!

Portable implies that something can be moved, but not how far or how easily. My first TV was a "portable," which meant that it featured handholds, yet the TV weighed about 25 pounds. *Cordless* implies a degree of freedom beyond portable, but still requires a charging station. Cordless devices typically have a range beyond which they fail to communicate. *Wireless* is at the top of the mobility heap. It implies freedom from wires but is also lightweight and tiny.

Your Android phone is truly wireless. Some devices don't even require a wire for charging. Even when you need a wire to charge the battery, once that process is complete, you can take your phone anywhere and use it wire-free. You can access the mobile network, a Wi-Fi network, and even wireless gizmos and peripherals. This chapter describes the degrees to which you can keep your phone wireless.

Wireless Networking Wizardry

You know that wireless networking has hit the big-time when you ask Santa Claus for a Wi-Fi router at Christmas. Such a thing would have been unheard of years ago, because back then routers were used primarily for woodworking. Today, wireless networking is what keeps gizmos like your Android phone connected to the Internet.

Using the mobile data network

You pay your cellular provider a handsome fee every month. The fee comes in two chunks: One chunk (the less expensive of the two) covers the telephone service; the second chunk delivers the data service, which is how your Android phone gets on the Internet. This system is the *mobile data network.*

Your phone is capable of accessing the mobile data network at various speeds. These current speeds and services are available:

> **4G LTE:** The fourth generation of wide-area data networks is the fastest and most popular network. Some providers may refer to this type of network as HSPA.

> **3G:** The third-generation mobile data network is available in locations that don't offer 4G LTE service.

> **1X:** The original mobile data network had no name, but is now called 1X. This service might be available when the other two have been obliterated by some moron with a backhoe.

Your phone always uses the best network available. So, whenever a 4G LTE network is within reach, that network is used for Internet communications. Otherwise, the 3G network is chosen, and then 1X as a form of last-ditch desperation.

>> Some phones feature a status icon that indicates the currently connected network.

>> The H+ status icon on some phones represents the HSPA mobile data network, which is equivalent to 4G LTE.

>> The Signal Strength icon might represent the mobile data network, but on some phones it refers only to the telephone service.

>> It's possible to make phone calls when the mobile data network is unavailable. In some remote locations, that's the only type of wireless service offered.

>> See Chapter 24 for information on how to monitor mobile data usage and avoid surcharges.

TIP

>> When both a mobile data network and Wi-Fi are available, your phone uses Wi-Fi for all Internet access. Therefore, I recommend connecting to, and using, a Wi-Fi network wherever possible.

Understanding Wi-Fi

The mobile data connection is nice and is available pretty much all over, but it costs you money. A better option, and one you should seek out whenever it's available, is *Wi-Fi*, or the same wireless networking standard used by computers for communicating with each other and the Internet.

Making Wi-Fi work on your Android phone requires two steps. First, you must activate the phone's Wi-Fi radio. The second step is connecting to a specific wireless network. The next two sections cover the details.

TECHNICAL
STUFF

Wi-Fi stands for *wireless fidelity*. It's brought to you by the numbers 802.11 and various letters of the alphabet too many to mention.

Activating Wi-Fi

Follow these steps to activate Wi-Fi on your Android phone:

1. Open the Settings app.

2. Choose Wi-Fi.

   On some Samsung phones, tap the Connections tab to locate the Wi-Fi item.

3. Ensure that the Wi-Fi Master Control icon is on.

To turn off Wi-Fi, repeat the steps in this section, but in Step 3 slide the master control to Off. Turning off Wi-Fi disconnects your phone from any wireless networks.

TIP

REMEMBER

» You can quickly peruse your phone's Wi-Fi settings by choosing the Wi-Fi quick setting. See Chapter 3 for more information on Quick Settings.

» It's perfectly okay to keep the phone's Wi-Fi radio on all the time. It is not a major drain on the battery.

» Using Wi-Fi to connect to the Internet doesn't incur data usage charges — unless you're using a metered connection. See the later section "Setting a metered Wi-Fi connection."

Connecting to a Wi-Fi network

After activating the Wi-Fi radio on your Android phone, you can connect to an available wireless network. Any available network you've previously connected to is automatically reconnected. Otherwise, you can connect to an available network by following these steps:

1. **Open the Settings app.**

2. **Choose Wi-Fi or Wireless & Networks.**

   On some Samsung phones, the Wi-Fi item is found on the Connections tab.

3. **Choose a wireless network from the list.**

   Available Wi-Fi networks appear on the screen, similar to what's shown in Figure 17-1. When no wireless networks are listed, you're sort of out of luck regarding wireless access from your current location.

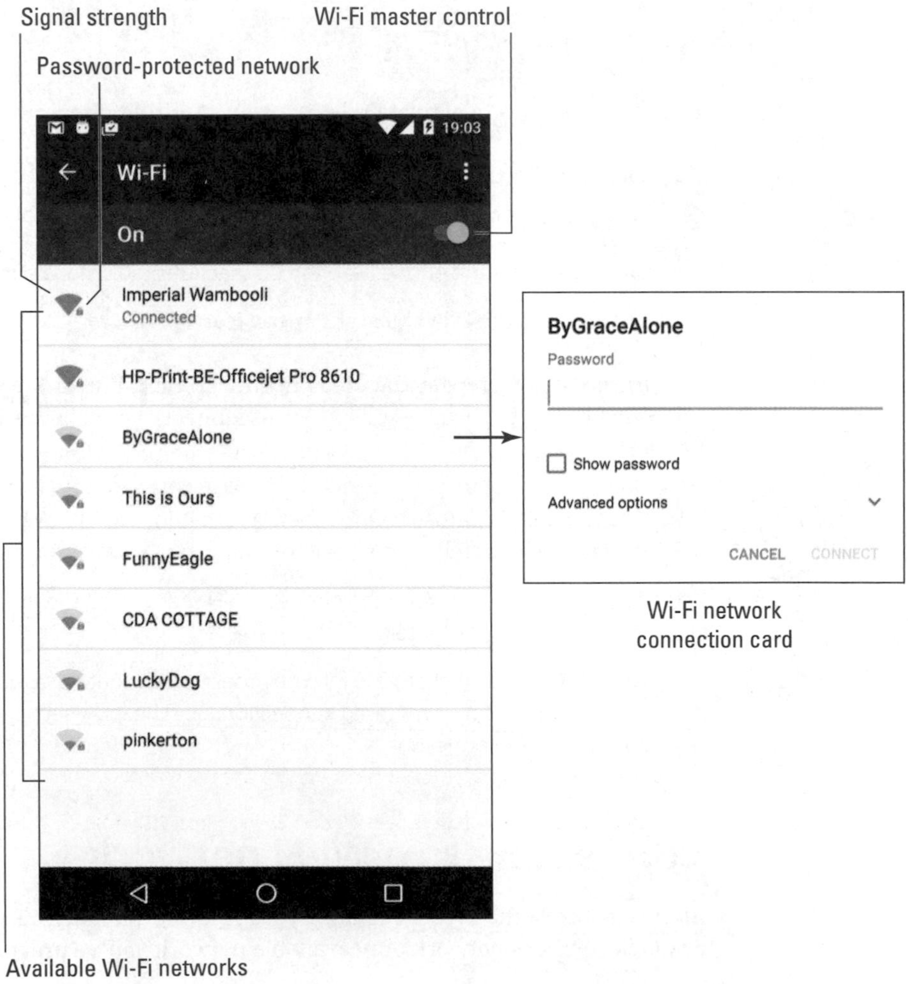

Signal strength   Wi-Fi master control

Password-protected network

ByGraceAlone

Password

☐ Show password

Advanced options   ⌄

      CANCEL CONNECT

Wi-Fi network connection card

FIGURE 17-1: Hunting down a wireless network.

Available Wi-Fi networks

4. **If prompted, type the network password.**

TIP

Tap the Show Password box so that you can see what you're typing. That helps when the password is intolerably long or complex.

5. **Tap CONNECT.**

The network is connected immediately. If not, try the password again.

When the phone is connected, you see the Wi-Fi status icon atop the touchscreen, looking similar to the icon shown in the margin. This icon indicates that the phone's Wi-Fi is on, connected and communicating with a Wi-Fi network.

>> Some public networks are open to anyone, but you have to use the phone's web browser app to find a login web page. Simply browse to any page on the Internet, and the login web page shows up. Follow the directions to get network access.

WARNING

>> Not every wireless network has a password. They should! I don't avoid connecting to any public network that lacks a password, but I don't use that network for shopping, banking, or any other secure online activity.

>> Unlike the mobile network, a Wi-Fi network's broadcast signal goes only so far. My advice is to use Wi-Fi when you plan to remain in one location for a while. If you wander too far away, your phone loses the signal and is disconnected

Connecting to a hidden Wi-Fi network

Some wireless networks don't broadcast their names, which adds security but also makes accessing them more difficult. In these cases, follow these steps to connect to the hidden Wi-Fi network:

1. **Open the Settings app and choose Wi-Fi.**

2. **Tap the Action Overflow.**

On Samsung phones, tap MORE.

3. **Choose Add Network or Add Wi-Fi Network.**

Some phones may use the Add (plus sign) icon to add a hidden Wi-Fi network.

4. **Type the network name into the Enter the SSID box.**

5. **Choose the security setting.**

6. **Type the password.**

The password may be optional. See the preceding section or my advice on using password-less public networks.

7. **Tap SAVE.**

   The network is connected automatically. If not, tap CONNECT.

Obtain the SSID, security, and password information from the girl with the pink hair and pierced lip who sold you coffee or from whoever is in charge of the wireless network at your location.

SSID stands for Service Set Identifier. It is not considered a valid acronym for use in Scrabble, despite its pitiful 5 points.

Connecting to a WPS router

Many Wi-Fi routers feature WPS, which stands for Wi-Fi Protected Setup. It's a network authorization system that's really simple and quite secure. If the wireless router features WPS, you can use it to quickly connect your phone to the network.

To make the WPS connection, follow these steps while standing near the Wi-Fi router.

1. **Open the Settings app and choose Wi-Fi.**

2. **If you can't find the WPS option or icon on the screen, tap the Action Overflow icon and choose Advanced. On Samsung phones, tap MORE.**

   Two WPS options are available, depending on how the router works. Some WPS routers are push-button activated; others list a PIN.

3. **Choose WPS Push Button or WPS Pin Entry, depending on the router.**

   If you choose the WPS Push Button, then push the button on the router labeled with the WPS icon, shown in the margin.

   If you choose the WPS Pin Entry, type the PIN shown on the phone's screen into the router.

Connection with the router may take a few minutes, so be patient. The good news is that, like all Wi-Fi networks, after the initial connection is established, the phone automatically connects in the future.

Setting a metered Wi-Fi connection

A *metered* Wi-Fi connection is one you pay for, either per minute or per megabyte or gigabyte of transferred data. Similar to the mobile data connection, when the phone accesses a metered Wi-Fi connection, you want to ensure that you don't exceed your data quota. To help you:

1. **Open the Settings app.**

2. **Choose Data Usage.**

   This item is located in the Wireless & Networks area, or on the Connection tab on certain Samsung phones.

3. **Tap the Action Overflow icon or MORE button.**

4. **Choose Network Restrictions.**

   This action might be titled Restrict Networks.

5. **By a listed Wi-Fi connection, set the master control to the On position.**

   Memorized Wi-Fi connections appear in the list.

When the Master Control icon is on (activated), the phone restricts data access over the Wi-Fi network. You will be warned whenever a large download or upload is attempted.

Happily, not many Wi-Fi networks require payment to access. If one does, set up the network as described in this chapter, and then follow the steps in this section to apply the metered restriction.

Share the Connection

Your Android phone need not jealously guard its mobile data connection. It's possible to share that Internet access in one of two ways. The first is to create a mobile *hotspot*, which allows any Wi-Fi–enabled gizmo to access the Internet via your phone. The second is a direct connection between your phone and another device, which is called *tethering*.

Creating a mobile hotspot

To share your phone's mobile data signal with other Wi-Fi devices in the vicinity, heed these steps:

1. **Open the Settings app.**

2. **Turn off the phone's Wi-Fi radio.**

   Why create a Wi-Fi hotspot when one is already available?

3. **Plug the phone into a power source.**

   The mobile hotspot feature draws a lot of power.

4. **Tap the Back navigation icon to return to the main Settings app screen.**

5. **Choose the More item.**

   You'll find the More item in the Wireless & Networks section of the Settings app. On some Samsung phones, locate the More item on the Connections tab, or you may find an item titled Mobile Hotspot and Tethering, in which case you can skip to Step 7.

6. **Choose Tethering & Portable Hotspot.**

   This item might be titled Mobile Hotspot.

7. **Slide the Portable Wi-Fi Hotspot master control to the On position.**

   On some phones, the item may be titled Mobile Hotspot. It might also feature a check box instead of a master control.

   After the hotspot is active, you need to confirm some of the settings.

8. **Choose Set Up Wi-Fi Hotspot.**

   On some Samsung phones, tap the MORE button and choose Configure Mobile Hotspot.

   Tap the fields on the Set Up Wi-Fi Hotspot card to assign a name, security level, and password.

9. **Tap the SAVE or OK button when you're done setting options on the Set Up Wi-Fi Hotspot card.**

The phone's Wi-Fi hotspot can be accessed by any other mobile device or computer with a Wi-Fi radio.

While the mobile hotspot is active, the Hotspot Active status icon appears, similar to the one shown in the margin.

To turn off the mobile hotspot, repeat the steps in this section, but disable the setting in Step 7.

>> Your phone's capability to create a mobile hotspot can be limited by the cellular provider. Some require an extra subscription (fee) before you can access this feature.

>> Some phones may feature a Mobile Hotspot or 4G Hotspot app. If so, use it instead of following the steps in this section.

>> The range for the mobile hotspot is about 30 feet. Items such as walls and tornadoes can interfere with the signal, rendering it much shorter.

>> The default Wi-Fi hotspot password is your phone number.

TIP

>> When you use the mobile hotspot, data usage fees apply. When a crowd of people are using the hotspot, a lot of data is consumed rather quickly.

>> Be sure to turn off the mobile hotspot when you're done using it.

Tethering the Internet connection

A more intimate way to share the phone's mobile data connection is to connect the phone directly to a computer and activate the tethering feature. Follow these steps to set up tethering:

1. **Use the USB cable to connect the phone to a computer or laptop.**

   I've had the best success with this operation when the computer is a PC running Windows.

2. **Open the Settings app and choose the More item, found in the Wireless & Networking area.**

   On some Samsung phones, choose the Mobile Hotspot and Tethering item, which may be found on the Connections tab.

3. **By the USB Tethering item, slide the master control to the On position.**

   On some phones, choose the Tethering item and then activate USB Tethering.

The other device should instantly recognize the phone as a "modem" with Internet access. Further configuration may be required, which depends on the computer using the tethered connection. For example, you may have to accept the installation of new software in Windows.

To end the connection, repeat Steps 2 and 3 in this section, but disable tethering in Step 3. Then you can disconnect the USB cable.

>> While tethering is active, the Tethering status icon may appear, similar to the one shown in the margin.

>> Sharing the mobile data connection incurs data usage charges against your cellular data plan. Be careful with your data usage when you're sharing a connection.

The Bluetooth Connection

Technology nerds have long had the desire to connect high-tech gizmos with one another. The Bluetooth standard was developed to sate this desire in a wireless way. Although Bluetooth is wireless *communication*, it's not the same as wireless networking. It's more about connecting peripheral devices, such as those wireless earpieces that are so popular among the golf shirt crowd.

Understanding Bluetooth

To make Bluetooth work on your phone, you need a Bluetooth peripheral, such as a wireless earpiece. The goal is to pair that gizmo with your phone. Here's how the operation works:

1. **Turn on the Bluetooth radio on both your phone and the peripheral.**

2. **Make the peripheral discoverable.**

   The peripheral broadcasts that it's available for dating other devices.

3. **On your phone, choose the peripheral from the list of Bluetooth devices.**

4. **If prompted, confirm the connection on the peripheral device.**

   For example, you may be asked to input a code or press a button.

5. **Use the Bluetooth peripheral.**

When you're done using the device, turn it off. Because the peripheral is paired with your phone, it's automatically reconnected the next time you turn it on — as long as the phone's Bluetooth radio is active.

 Bluetooth devices are marked with the Bluetooth logo, shown in the margin. It's your assurance that the gizmo can work with other Bluetooth devices.

Activating Bluetooth on the phone

You must turn on the phone's Bluetooth radio before you can use one of those Borg-earpiece implants and join the ranks of the walking connected. Assimilate these steps:

1. **Open the Settings app.**

2. **Choose Bluetooth.**

   On some Samsung phones, you'll find the Bluetooth item on the Connections tab.

3. **Ensure that the Bluetooth master control is set to the On position.**

   Slide the icon to the right to activate.

When Bluetooth is on, the Bluetooth status icon appears. It uses the Bluetooth logo, shown in the margin.

To turn off Bluetooth, repeat the steps in this section, but slide the master control to the Off position in Step 3.

The Bluetooth quick setting can also be used to activate and deactivate the phone's Bluetooth radio. See Chapter 3 for information on Quick Settings.

Pairing with a Bluetooth peripheral

To make the Bluetooth connection between your phone and some other gizmo, such as a Bluetooth headset, follow these steps:

1. **Ensure that phone's Bluetooth radio is on.**

   Refer to the preceding section.

2. **Make the Bluetooth peripheral discoverable.**

   Turn on the gizmo and ensure that its Bluetooth radio is on. If the device has separate power and Bluetooth switches, press the Bluetooth button or take whatever action is necessary to make the peripheral discoverable.

3. **On the phone, open the Settings app and choose Bluetooth.**

   The Bluetooth screen shows already paired and available peripherals, similar to what's shown in Figure 17-2. If not, tap the Refresh button or Scan button, or tap the Action Overflow to look for similar actions.

4. **Choose the Bluetooth peripheral from the list.**

5. **If necessary, type the device's passcode or otherwise acknowledge the connection.**

   For example, press the button on the Bluetooth earpiece to complete the pairing.

After pairing, you can begin using the device.

Connected devices appear on the Bluetooth screen, under the heading Paired Devices, such as the Nica earpiece. (Refer to Figure 17-2.)

>> To break the connection, you can either turn off the gizmo or disable the Bluetooth radio on your phone. Because the devices are paired, when you turn on Bluetooth and reactivate the device, the connection is reestablished.

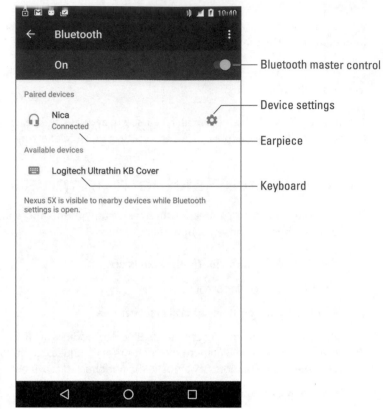

Bluetooth master control

Device settings

Earpiece

Keyboard

Nexus 5X is visible to nearby devices while Bluetooth settings is open.

FIGURE 17-2:
Finding Bluetooth gizmos.

>> It's rare to unpair a device. Should you need to, visit the Bluetooth screen (refer to Figure 17-2) and tap the Settings icon by the device's entry. Choose the Unpair action or tap the OK button to terminate the pairing.

>> Unpair only the devices that you don't plan to use again. Otherwise, just turn off the Bluetooth device when you're done using it.

REMEMBER

>> The Bluetooth radio consumes a lot of power. Don't forget to turn off the device, especially a battery-powered one, when you're no longer using it with your phone.

Chapter 18

Share and Store

At some point, you may treasure the items inside your phone so much that you want to access them elsewhere. Or the opposite may be true, and you desire to add to your phone's inventory photos, music, or similar items. Either way, the goal is to transfer information. It seems like a simple operation, but as with most things high-tech, it's a wee bit more complex than you would think.

From Here to There

A variety of useful and interesting methods exist for transferring files between your phone and other devices, such as a computer. None of these methods involves using tweezers or magnets.

>> Getting a file into your phone is no guarantee that you can do productive things with it. Specifically, don't expect to be able to read an eBook file you've copied from elsewhere.

>> Media you've obtained from Google Play, including music, eBooks, movies, and TV shows, is accessible from any device that has Internet access. Likewise, your phone's photo and video library may be accessed remotely, as described elsewhere in this book.

» A good understanding of basic file operations is necessary for successful file transfers between a computer and your phone. Knowing basic procedures such as copy, move, rename, and delete is an important part of the process. Understand how a folder works. The good news is that you don't need to manually calculate a 64-bit cyclical redundancy check on the data, nor do you need to know what a parity bit is.

Sharing files on the cloud

The simplest and most effective way to swap files between your Android phone and just about any other device is to use cloud storage. As a Google account holder, you can use the Drive app to access your Google Drive storage. All files saved to your Google Drive are synchronized instantly with other devices that access that storage. Change a file in a Google Drive folder, and that file is instantly updated on your phone, computer — everywhere.

To move an item from a computer to the phone, first ensure that the Google Drive program is installed on the computer: Visit `drive.google.com` and follow the directions to install the program.

After the Google Drive software is installed on the computer, copy the file(s) you want to transfer to your computer's Google Drive folder. In mere seconds, the file is accessible on your phone, courtesy of the Drive app.

To copy an item from the phone to your computer, follow these steps:

1. **Locate the item you want to send to the computer.**

It can be a picture, movie, or web page or just about any file on the phone.

 2. **Tap the Share icon, or tap the Action Overflow and look for a Share action.**

If you don't see the Share icon, the item you're viewing cannot be copied to the Google Drive.

3. **Choose Save to Drive.**

A card appears, similar to the one shown in Figure 18-1.

4. **Edit the filename or change the folder, if you so desire.**

I typically change the name to something memorable. Also, because I'm organized, I tap the Folder action bar to choose a specific Google Drive folder on which to save the item.

5. **Tap the SAVE button.**

Name selected for editing

Choose folder

FIGURE 18-1:
Sharing a
file to
Google Drive.

Other cloud storage apps include the popular Dropbox, Microsoft's OneDrive, the Amazon Cloud, and more. You can use those cloud storage locations just as easily as Google Drive; choose the proper app in Step 3.

REMEMBER

>> You can obtain the Dropbox app from Google Play, but you also need to get a Dropbox account and, probably, get the Dropbox program for the computer. Visit www.dropbox.com to get started.

>> Cloud storage apps are free, and you can use a token amount of storage at no charge.

Using the USB cable to transfer files

A more direct way to transfer files between a computer and your phone is to physically connect the two devices. Because this is an ancient and traditional method of file exchange, it's also a tad bit more complex and confusing than using cloud storage to meet the same end.

To perform the direct-connection file swap, obey these steps on a computer running Windows:

1. **Use the USB cable that came with the phone to connect the phone and a computer.**

2. **If your phone features a secure screen lock, unlock the phone.**

The computer cannot access the phone's storage while the phone is locked with a PIN or password or another secure screen lock. See Chapter 21 for information.

3. **View the phone's storage on the computer.**

If the AutoPlay notification appears, which happens most of the time, choose Open Folder to View Files. Otherwise, you may automatically see a folder window that shows the phone's files and folders. If not, press Win+E to open a File Explorer window and choose the phone's icon from those listed on the left side of the window, as illustrated in Figure 18-2.

4. **Open the folder to which or from which you want to copy files.**

Arrange both folder windows, computer and phone, on the computer screen, similar to what's shown in Figure 18-2.

5. **Drag the file icon(s) from one folder window to the other.**

Dragging the file copies it.

6. **Close the folder windows when you're done copying files.**

7. **Disconnect the USB cable from the computer.**

TIP

If you want to be specific, drag the file from the computer to the phone's Download folder. (Refer to Figure 18-2.) On the PC, drag files to the My Documents folder.

On the odd chance that you use an Android phone with a Macintosh, you must obtain special software to perform a USB cable file transfer. Obtain the Android File Transfer program from this website:

```
www.android.com/filetransfer
```

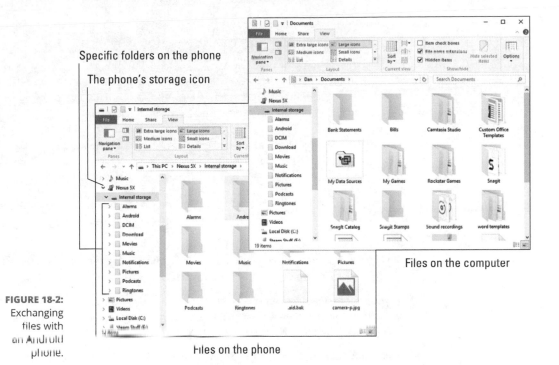

Specific folders on the phone

The phone's storage icon

Files on the computer

FIGURE 18-2:
Exchanging files with an Android phone.

Files on the phone

Install the software. Run it. From that point on, whenever you connect your Android phone to the Macintosh, a special window appears, similar to the one shown in Figure 18-3. It lists the phone's folders and files. Use that window for file management, as covered in this section.

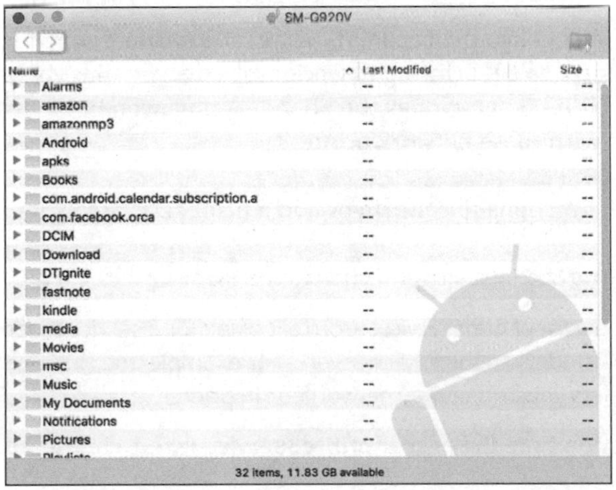

FIGURE 18-3:
The Android File Transfer program.

MEDIA CARD TRANSFER

If your phone features removable storage, you can use it to transfer files. Remove the microSD card from the phone and insert it into a computer. From that point, the computer can read the files just as they can be read from any media card.

See Chapter 1 for details on how to remove the microSD card from your phone. You can't just yank out the thing! You also need a microSD adapter to insert the card into a media reader on the computer. Or, you can get a microSD card USB thumb drive adapter.

>> If you're transferring media, such as music or photos, use the Windows Media Player program. See Chapter 14 for information.

>> If your phone features removable storage, you'll see two storage items appear on your computer. One is for the phone's main storage; the other represents the phone's media card.

>> Files you've downloaded on the phone are stored in the Download folder.

>> Pictures and videos on the phone are stored in the DCIM/Camera folder.

>> Music on the phone is stored in the Music folder, organized by artist.

Adding a print service

Android phones have the capability to print information, which is another form of file transfer. To get printing to work properly, you must have printers available on the same Wi-Fi network as the phone — plus, you must install a print service compatible with those network printers.

Heed these directions to confirm or add a print service to your phone:

1. Open the Settings app and choose the Printing category.

You see a list of print services. You're looking for a service that matches the printer models on the Wi-Fi network. For example, the HP Print Service Plugin, which lets you print to any networked HP printer.

2. If you don't see the proper printer service listed, choose Add Service.

The Play Store app opens, listing available printing services.

3. **Choose and install a print service.**

   For example, if you use Canon printers, choose and install the Canon Print service. Refer to Chapter 16 for details on installing apps on your phone.

After the service is installed, or if you confirm that the service is available, you can print from your Android phone. Keep reading in the next section.

Printing

When printing services are available (refer to the preceding section), printing on an Android phone works similarly to printing on a computer. Follow these steps:

1. **View the material you want to print.**

   You can print a web page, photo, map, or any number of items.

2. **Tap the Action Overflow.**

   Some Samsung phones use the MORE button in place of the Action Overflow icon.

3. **Choose Print.**

   If you don't see the Print action, your phone lacks this feature or the material cannot be printed.

4. **Select a printer.**

   The current printer is shown on the action bar, as illustrated in Figure 18-4. If that's not the printer you want to use, tap the action bar to view printers available on the currently connected Wi-Fi network.

5. **To change any print settings, tap the Show More Details chevron.**

   The items presented let you set which pages you want to print, change the number of copies, and make other common printer settings.

6. **Tap the Print button.**

   The item prints.

REMEMBER

In addition to being on the same Wi-Fi network as the phone, the printer must be on, stocked with the proper paper, and ready to print.

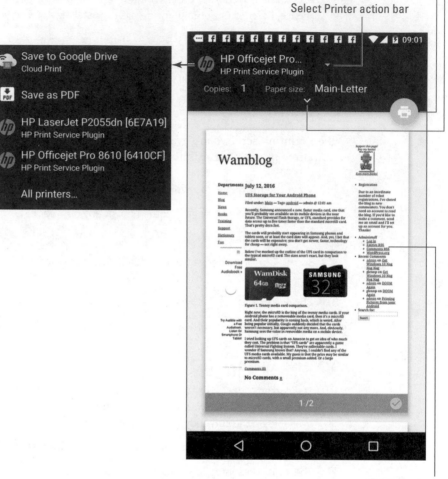

Show more details

Print

Select Printer action bar

Save to Google Drive
Cloud Print

Save as PDF

HP LaserJet P2055dn [6E7A19]
HP Print Service Plugin

HP Officejet Pro 8610 [6410CF]
HP Print Service Plugin

All printers...

HP Officejet Pro...
HP Print Service Plugin

Copies: 1   Paper size: Main-Letter

Print preview

FIGURE 18-4:
Android phone
printing.

Streaming your own media

When you desire to watch movies, look at your photos, or listen to music on a large-format screen, it's time to *screencast:* This technology takes the media presented on your phone (video or music) and streams it to an HDTV, a monitor, or another compatible device.

To successfully stream media, you need a casting gizmo attached to the output device. Google prefers that you use the Chromecast dongle, which works naturally with your Android phone. When that step is completed, streaming your media works like this:

1. **Open the app that plays the media you want to watch or listen to.**

   Compatible apps include Play Music, Play Movies & TV, YouTube, Netflix, and so on.

2. **Tune the HDTV or monitor to the proper HDMI input.**

   For example, if a Chromecast dongle is installed on HDMI Input 4, switch the TV to that input. The casting dongle must be awake and active.

3. **Tap the Chromecast icon.**

   The icon appears similar to the one shown in the margin. If you don't see this icon, either the Chromecast dongle isn't awake or the media cannot be cast to another device.

4. **Choose an output device from the list.**

   The phone begins to send the media to the other device.

You can still use your phone while it's casting. The app on the screen may offer you controls, such as Play and Pause, or it might display additional information about the media.

To stop streaming, tap the Chromecast icon again, and then tap the DISCONNECT button.

Phone Storage Mysteries

Somewhere deep in your phone's bosom lies a storage device. It's like the hard drive in a computer: The thing can't be removed, but that's not the point. The point is that the storage is used for your apps, music, videos, pictures, and a host of other information stored on the phone.

>> Android phones come with 8GB, 16GB, 32GB, or more of internal storage.

>> Removable storage in the form of a microSD card is available on some phones. The microSD card capacity varies between 8GB and 64GB or greater.

>> GB stands for a gigabyte, or 1 billion bytes (characters) of storage. A typical 2-hour movie occupies about 4GB of storage. Most things you store on the phone don't occupy that much space. In fact, media stored on the cloud occupies hardly any storage space. Refer to Chapter 16 for details on how media is stored on the cloud and can be downloaded to your phone.

TECHNICAL STUFF

Reviewing storage stats

To see how much storage space is available on your phone, follow these steps:

1. **Open the Settings app.**

2. **Choose Storage & USB or Storage.**

On some Samsung phones, you'll find the Storage item on the Settings app's General tab.

A typical storage screen is shown in Figure 18-5. It details information about storage space on the phone's internal storage and, if available, on the microSD card.

If your phone has external storage, look for the SD Card category at the bottom of the storage screen (not shown in Figure 18-5).

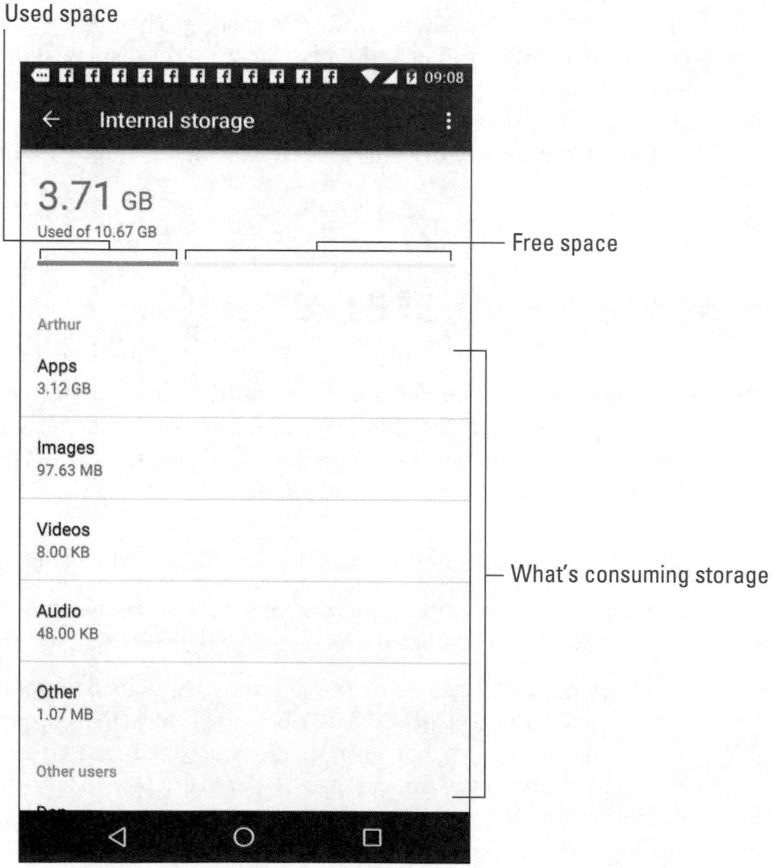

Used space

Free space

What's consuming storage

FIGURE 18-5: Phone storage information.

Tap a category on the storage screen to view details on how the storage is used or to launch an associated app. For example, tap Apps to view a list of running apps. If you tap the Videos item, you see a list of videos stored on the phone.

REMEMBER

>> Videos, music, and pictures (in that order) consume the most storage space on your phone.

>> Media that you download to the phone from the Google cloud, such as movies or videos, are tracked in the Apps category.

TECHNICAL STUFF

>> Don't bemoan the Total Space value being far less than the phone's storage capacity. For example, in Figure 18-5, a 16GB phone shows only 10.67 GB total storage. The missing space is considered overhead, as are several gigabytes taken by the government for tax purposes.

Managing files

TECHNICAL STUFF

You probably didn't get an Android phone because you enjoy managing files on a computer and wanted another gizmo to hone your skills. Even so, you can practice the same type of file manipulation on a phone as you would on a computer. Is there a need to do so? Of course not! But if you want to get dirty with files, you can.

To view files and folders, attempt these steps:

1. **Open the Settings app and choose Storage & USB.**

   This item might be titled Storage.

2. **Swipe to the bottom of the screen and choose Explore.**

   You see folders and files stored on your phone.

Some Android phones come with a file management app. It's called My Files or File Manager, and it's a traditional type of file manager, which means that if you detest managing files on your computer, you'll enjoy the same pain and frustration on your phone.

TIP

>> If you simply want to peruse files you've downloaded from the Internet, open the Downloads app, found in the apps drawer. Refer to Chapter 10.

>> When your phone lacks a file management app, you can swiftly obtain one. An abundance of file management apps are available from Google Play. See Chapter 16.

Unmounting the microSD card

The microSD card provides removable storage on a handful of Android phones. When the phone is turned off, you can insert or remove the microSD card at will; directions are provided in Chapter 1. The microSD card can also be removed when the phone is turned on, but you must first unmount the card. Obey these steps:

1. **Open the Settings app and choose Storage & USB or Storage.**

   On some Samsung phones, tap the General tab to locate the Storage item.

2. **Choose Unmount SD Card.**

   This item is found near the bottom of the screen. If not, your phone lacks the ability to host a removable microSD card.

3. **Ignore the warning and tap the OK button.**

4. **When you see the Mount SD Card action, it's safe to remove the microSD card.**

It's important that you follow these steps to safely remove the microSD card. If you don't and you simply pop out the card, it can damage the card and lose information.

REMEMBER

You can insert a microSD card at any time. See Chapter 1 for details.

Formatting microSD storage

Your Android phone is designed to instantly recognize a microSD card when it's inserted. If it doesn't, you can attempt to format the card to see whether that fixes the problem. Follow these steps to determine whether a format is in order:

1. **Open the Settings app and choose the Storage & USB item.**

2. **Choose Format SD Card.**

3. **Tap the Format SD Card button.**

WARNING

   All data on the microSD card is erased by the formatting process.

4. **Tap the DELETE ALL button to confirm.**

   Not every phone displays the DELETE ALL confirmation button.

The microSD card is unmounted, formatted, and then mounted again and made ready for use.

Chapter 19

The Apps-and-Widgets Chapter

Beyond making phone calls, your Android phone is host to an abundance of apps. The app launchers, as well as their widget cousins, adorn your phone's Home screen, which helps make the device more usable. Your goal is to keep these apps and widgets accessible without overwhelming the Home screen.

Apps and Widgets on the Home Screen

Your phone came with app launchers and widgets affixed to the Home screen. You can keep those launchers, add your own, remove some, and change their locations. And when the Home screen brims with launchers like an overflowing teacup, you can build folders to further organize and keep your apps neat and tidy.

REMEMBER

The app icons on the Home screen are called *launchers.*

Adding launchers to the Home screen

You need not live with the unbearable proposition that you're stuck with only the apps that come preset on the Home screen. Nope — you're free to add your own apps. Just follow these steps:

1. **Visit the Home screen page on which you want to stick the launcher.**

The screen must have room for the icon. If not, swipe the screen left or right to hunt down a page.

2. **Tap the Apps icon to display the Apps drawer.**

3. **Long-press the app you want to add to the Home screen.**

After a moment, the Home screen page you chose in Step 1 appears, similar to the one shown in Figure 19-1.

FIGURE 19-1:
Placing a
launcher
icon on the
Home screen.

Drag the icon to a
position on the
Home screen

Favorites tray

Apps icon

4. **Drag the app to a position on the Home screen.**

   Launchers are aligned to a grid. Other launchers may wiggle and jiggle as you find a spot. That's okay.

5. **Lift your finger to place the app.**

The app hasn't moved: What you see is a copy or shortcut. You can still find the app in the Apps drawer, but now the app is also available — more conveniently — on the Home screen.

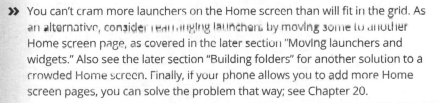

>> Any new apps you obtain from the Play Store are automatically affixed to the Home screen. If you prefer not to have the new app's launcher on the Home screen, you can remove it. See the later section "Evicting items from the Home screen."

>> Not every app needs a launcher icon on the Home screen. I recommend placing only those apps you use most frequently.

>> You can't cram more launchers on the Home screen than will fit in the grid. As an alternative, consider rearranging launchers by moving some to another Home screen page, as covered in the later section "Moving launchers and widgets." Also see the later section "Building folders" for another solution to a crowded Home screen. Finally, if your phone allows you to add more Home screen pages, you can solve the problem that way; see Chapter 20.

Putting a launcher on the favorites tray

The line of launchers at the bottom of the Home screen is called the *favorites tray*. It includes the Apps icon, as well as other launchers you place there, such as the Phone or Dialer, or any app you use most often. The favorites tray launchers appear at the bottom of every Home screen page.

Launchers are added to the favorites tray in one of two ways:

>> **Drag a launcher off the favorites tray,** either to the Home screen or to the Remove or Delete icon. You can then add another launcher to replace the one you removed.

>> **Drag a launcher from the Home screen to the favorites tray,** in which case any existing launcher swaps places with the one already there.

Of these two methods, the second one may not work on all phones. In fact, the second method may create a folder in the favorites tray, which is probably not what you want.

See the later section "Building folders" for information on creating and using Home screen folders, which can also dwell on the favorites tray.

The best icons to place on the Home screen are those that show pending items, notifications, or updates, similar to the icon shown in the margin. These icons are also ideal to place in the favorites tray, as covered in the next section.

Slapping down widgets

The Home screen is the place where you can find widgets, or tiny, interactive information windows. A widget often provides a gateway to a specific app or displays information such as status updates, the name of the song that's playing, or the current weather.

Two common methods are available to add widgets to the Home screen. The stock Android way is to view the Widgets drawer. An older method involved accessing widgets from a tab on the Apps drawer. Both methods are covered in these steps:

1. **Switch to a Home screen page that has enough room for the new widget.**

   Widgets come in a variety of sizes. The size is measured by launcher dimensions: A 1-x-1 widget occupies the same space as a launcher icon. A 2-x-2 widget is twice as tall and twice as wide as a launcher icon.

2. **Access the widgets.**

   On newer phones, long-press a blank part of the Home screen. Tap the Widgets icon.

   If you don't see the Widgets icon, tap the Back navigation icon, and then tap the Apps icon to view the apps drawer. Tap the Widgets tab atop the screen.

3. **Long-press the widget you want to add.**

   Swipe through the pages to find widgets, which are listed alphabetically and show preview images, just like apps. Also shown are the widget's dimensions.

4. **Drag the widget to the Home screen.**

   As you drag the widget, existing launchers and widgets jiggle to make room.

5. **Lift your finger.**

   If the widget grows a border, it can be resized. See the next section.

You may see a prompt for additional information after adding certain widgets — for example, a location for a weather widget or a contact name for a contact widget.

- » The variety of available widgets depends on the apps installed. Some apps come with widgets; some don't. Some widgets are independent of any app.

- » Just like launchers, widgets can be moved and rearranged on the Home screen. See the later section "Moving launchers and widgets," for obtaining the proper feng shui.

- » Also see the later section "Evicting items from the Home screen," for information on unsticking a widget from the Home screen.

Resizing a widget

Some widgets are of a fixed size, but many can be resized. You can resize a widget immediately after plopping it down on the Home screen, or at any time, really. The secret is to long-press the widget: If it grows a box, as shown in Figure 19-2, you can change the widget's dimensions.

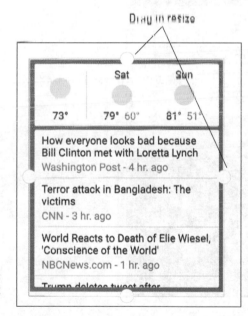

Drag to resize

FIGURE 19-2:
Resizing
a widget.

To resize, drag one of the resizing dots in or out, as illustrated in Figure 19-2. Tap elsewhere on the touchscreen when you're done resizing.

Moving launchers and widgets

Redecorating your Android phone's Home screen is a lot easier than trying to move a couch from one room in your house to another. And you don't have to bribe anyone with pizza to make it happen: Simply long-press the item, as illustrated in Figure 19-3.

Long-press to "lift"

Remove from Home screen   Uninstall app

FIGURE 19-3:
Moving
an item on the
Home screen.

Drag to another page   Drag to another page

Drag the launcher, widget, or folder to another position on the Home screen. If you drag to the far left or far right of the screen, the item is sent to another Home screen page.

Evicting items from the Home screen

When your rearranging plans include the removal of a launcher, widget, or folder, drag that item up to the top of the Home screen page, where the Remove icon lurks. The Remove icon can appear as shown in the margin, or it may look as shown in Figure 19-3. When the item hovers over that part of the screen, lift your finger.

REMEMBER

>> Feel no guilt or regret about removing anything from the Home screen, especially those launchers and widgets preinstalled by the phone's manufacturer or the cellular provider. If you don't use something on the Home screen, get rid of it.

>> Removing a launcher from the Home screen doesn't uninstall the app. Well, unless you drag it to the Uninstall icon. See the later section "Uninstalling apps."

Building folders

To further organize the Home screen, or to deal with an overly crowded Home screen, consider gathering similar apps into folders. A folder is simply a single icon that contains a host of launchers. For example, I have a Listen folder that contains all my streaming-music apps. Your phone may have come with a Google folder, which contains all of Google's various apps.

The stock Android method to create a folder is to drag one launcher over another on the Home screen. Be quick! When the two launchers get close, a circle encloses them, which provides visual feedback that a folder is created.

Some phones require that you first create a folder: Drag a launcher to a Folder icon or Create Folder icon. Or, you may have to first tap the Create Folder icon, name the folder, and then add launchers.

Figure 19-4 illustrates different ways that folders appear on various phones.

FIGURE 19-4:
Folder icon
varieties.

Google   **Google**   **Google**

To use a folder, tap its icon. The folder opens, displaying its contents. Tap a launcher to start the associated app. Or if you don't find what you want, tap the Back navigation icon to close the folder.

» To add more app launchers to the folder, drag in their icons. Some folders may feature a plus (+) or add icon. In that case, tap the icon to add more launchers.

» Folders are managed just like other items on the Home screen. You can rearrange folder icons and remove them just like launchers and widgets. Removing a folder doesn't uninstall its enclosed apps.

» To change a folder's name, open the folder and then tap the current name. Use the onscreen keyboard to type a new name.

» To remove a launcher from a folder, open the folder and drag out the icon. When the second-to-last icon is dragged out of a folder, the folder is removed. If not, drag the last icon out, and then remove the folder as described in the earlier section "Evicting items from the Home screen."

Manage Those Apps

The apps you install on your Android phone originate from Google Play. And that's where you can return for app management. The task includes reviewing apps you've downloaded, updating apps, and removing apps that you no longer want or that you severely hate.

Reviewing your apps

To peruse the apps obtained from Google Play, follow these steps:

1. **Open the Play Store app.**

2. **Ensure that you're at the top level in the app.**

   Tap the Back navigation icon until you see the main screen.

3. **Tap the Side Menu icon.**

4. **Choose My Apps & Games from the navigation drawer.**

   If you don't see the My Apps & Games item, choose Apps & Games and repeat Steps 3 and 4.

5. **Peruse your apps.**

Apps are presented on two tabs, Installed and All, as shown in Figure 19-5. Apps currently installed on your phone are listed on the Installed tab. The All tab lists every app you've ever obtained from Google Play, whether or not it's currently installed on the phone.

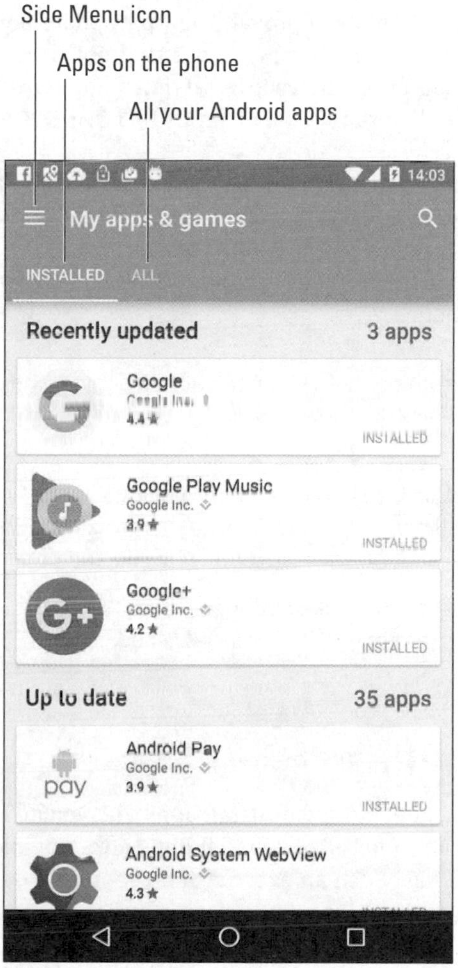

Side Menu icon

Apps on the phone

All your Android apps

FIGURE 19-5:
Apps on
your phone.

>> Tap an app in the list to view its details. Some of the options and settings on the app's details screen are discussed elsewhere in this chapter.

>> Uninstalled apps remain on the All tab's list because you did, at one time, download the app. To reinstall an app (and without paying a second time for paid apps), choose the app from the All list and tap the Install button.

Updating apps

New versions, or *updates*, of apps happen all the time. It's nothing you need to worry about, because most app updates are installed automatically. Occasionally, however, you must tend to manual updates.

 When apps are in need of update approval, the App Update notification appears, similar to the one shown in the margin. Choose that notification, or visit the My Apps screen as described in the preceding section (and shown in Figure 19-5). Manual updates are listed on the screen by an UPDATE or UPDATE ALL button.

For some app updates, you must accept permissions. Review the phone features accessed, and then tap the Accept button when prompted.

You can view the app update process in the Play Store app or go off and do something else with your phone.

>> Most apps update automatically; you need not do a thing. The updates generate a notification icon, shown in the margin. Feel free to dismiss the notification.

>> An Internet connection is required in order to update apps. If possible, try to use Wi-Fi so that you don't incur any data surcharges on your cellular bill. Android apps aren't super-huge in size, but why take the risk?

>> If the Internet connection is broken during an update, the updates automatically continue when the connection is reestablished.

TECHNICAL
STUFF

Uninstalling apps

It doesn't happen often, but you can uninstall apps that you no longer use or those that vex you incessantly. I typically uninstall apps after downloading a clutch of them to find one that I like. Such an exercise isn't required, but it helps keep me sane, which is an item somewhere near the top of my priority list.

When the need arises, remove an app by following these steps:

1. **Open the Play Store app.**

2. **Choose My Apps & Games from the navigation drawer.**

3. **Tap the Installed tab.**

4. **Choose the app that offends you.**

5. **Tap the UNINSTALL button.**

6. **Tap the OK button to confirm.**

   The app is removed.

The app continues to appear on the All list even after it's been removed. That's because you downloaded it once, but it doesn't mean that the app is currently installed on your phone.

WARNING

>> You can always reinstall paid apps that you've uninstalled. You aren't charged twice for doing so.

>> You can't remove apps that are preinstalled on the phone by either the phone's manufacturer or cellular provider. I'm sure there's a technical way to uninstall the apps, but seriously: Just don't use the apps if you want to remove them and discover that you can't.

Choosing a default app

While tapping links or items or attempting to view a download, you may find yourself confronted with the Open With card, which looks similar to the one shown in Figure 19-6. This card might be titled Complete Action Using on some phones.

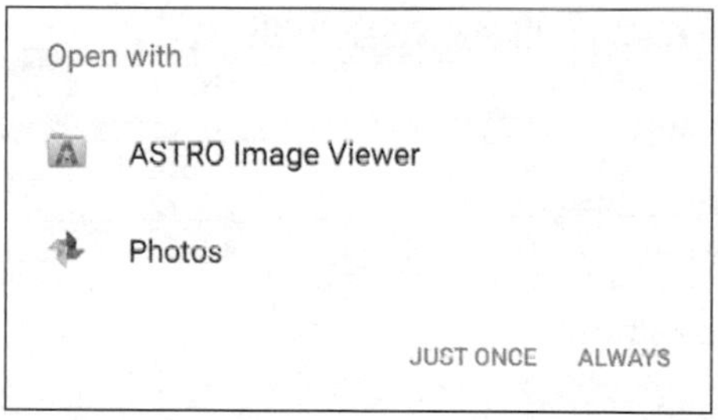

```
Open with

A    ASTRO Image Viewer

     Photos


                    JUST ONCE    ALWAYS
```

FIGURE 19-6:
An Open
With card.

Your duty is to choose from one or more apps to complete an action. Then you're given the choice of the options Just Once or Always.

>> When you choose Always, the chosen app always completes the action; you'll never see the Open With card again for that specific task.

>> When you choose Just Once, you see the Open With card again.

My advice is to choose Just Once until you get sick of seeing the Open With card. At that point, choose Always.

The fear, of course, is that you'll make a mistake. Keep reading in the next section.

Clearing default apps

Fret not, gentle reader. The action you chose for the Open With card can be undone. For example, if you select the ASTRO Image Viewer app, shown in Figure 19-6, you can undo that choice by following these steps:

1. **Open the Settings app.**
2. **Choose Apps.**
3. **Select the app that always opens.**

    This is the tough step because you must remember which app you chose to "Always" open.

4. **Choose the item Open by Default.**

    If the item shows the text *Some Defaults Set,* you're on the right track. Apps that haven't been chosen as the default show the text *No Defaults Set.*

5. **Tap the CLEAR DEFAULTS button.**

The steps on some Samsung phones are different:

1. **Open the Settings app.**
2. **Choose Applications.**

    Some Samsung phones may feature an Applications tab, in which case you choose that tab to continue.

3. **Choose Default Applications.**
4. **Choose Set as Default.**

    You see a list of installed apps. If the text *Set As Default* appears below an app, it's been chosen as a default app.

5. **Tap to open the default app.**
6. **Tap the CLEAR DEFAULTS button.**

After you clear the defaults for an app, you see the Open With card again. The next time you see it, however, make a better choice.

Chapter 20

Customize and Configure

C ustomizing your Android phone doesn't involve sprucing up the phone's case, so put away that Bedazzler™. The kind of customization this chapter refers to involves fine-tuning the way the Android operating system presents itself. You can modify the Home screen, adjust the display, customize the keyboard, and change sounds. The goal is to truly make your phone your own.

It's Your Home Screen

The typical Android Home screen sports anywhere from three to nine pages and a specific background, or wallpaper, preset by the phone's manufacturer. Or, if the phone has a preset wallpaper touting your cellular provider's new 4G LTE service, I'd guess that they provided the wallpaper. No matter how the Home screen is presented, it's yours to change at your whim.

Finding the Home screen menu

To control the Home screen, you must access its menu: Long-press a blank part of the Home screen; do not long-press a launcher or widget. The menu actions may present themselves as icons or as a list of actions on a card. Both presentations are shown in Figure 20-1.

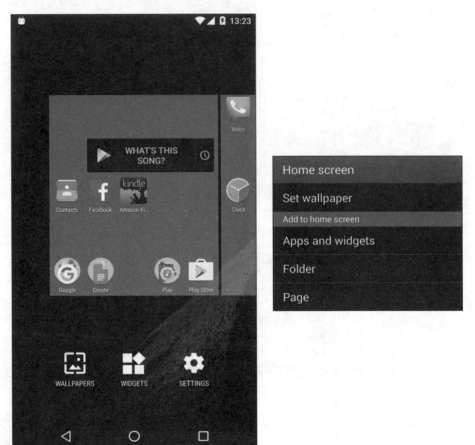

FIGURE 20-1:
The Home
screen menus.

Home screen management icons   Home screen management card

The Home screen menu includes some or all of the following actions:

Wallpapers: Change the background image on the Home screen. This item might be titled Wallpaper or Set Wallpaper.

Widgets: Choose a new widget to affix to the Home screen. This item might be titled Apps and Widgets.

Settings: Start the Settings app.

Folder: Create folders for multiple apps on the Home screen.

Page: Add, remove, or manage multiple Home screen pages.

More actions might be available as well, depending on the phone. The most common items are shown at the bottom left in Figure 20-1.

Changing wallpaper

The Home screen background can be draped with two types of wallpaper: traditional and live. Traditional wallpaper can be any image, such as a picture you've taken or an image provided by the phone's manufacturer. Live wallpaper is animated or interactive.

To hang new wallpaper on the Home screen, obey these steps:

1. **Long-press the Home screen.**

2. **Choose Wallpapers or Set Wallpaper**

   Refer to Figure 20-1 to see how the actions could be presented.

3. **Tap a wallpaper to see a preview.**

   Swipe the list left or right to peruse your options. You see the previous wallpaper images plus those provided by the phone's manufacturer. On the far right, you'll find the live wallpapers.

4. **Tap the Set Wallpaper button to confirm your choice.**

   The new wallpaper takes over the Home screen.

Some phones may prompt you to set the Home screen wallpaper, lock screen, or both.

To select an image stored in the phone, or one you've taken by using the phone's camera, choose the item My Photos or From Gallery. Select an image. You may be prompted to crop the image so that it fits properly on the Home screen.

You might be prompted to choose a wallpaper type before you can view samples. The list includes Wallpapers (preset images), Live Wallpapers, and Gallery or Photos, which represents images you've taken.

TIP

>> Be careful how you crop the wallpaper image when you choose one of your own photos. Zoom out (pinch your fingers on the touchscreen) to ensure that the entire image is cropped properly for the phone's horizontal and vertical orientations.

>> Live wallpapers can be obtained from Google Play. See Chapter 16. In particular, check out the Zedge app.

>> See Chapter 13 for more information about the images stored on your phone, including information on how to crop an image.

Managing Home screen pages

The number of pages on the Home screen isn't fixed. You can add pages. You can remove pages. You can even rearrange pages. This feature might not be available to all Android phones and, sadly, it's not implemented in exactly the same way on every phone.

The stock Android method of adding a Home screen page is to drag a launcher left or right, just as though you were positioning that item on another Home screen page. When a page to the left or right doesn't exist, the phone automatically adds a new, blank page.

Likewise, to remove a page, simply drag off the last launcher. The page vanishes.

Other phones may be more specific in how pages are added. For example, you might be able to choose a Page command from the Home Screen menu. (Refer to Figure 20-1.)

Samsung phones feature a Home screen page overview, shown in Figure 20-2. To edit Home screen pages, pinch the Home screen. On older Samsung phones, long-press the Home screen and choose Edit Page. You can then manage Home screen pages as illustrated in the figure.

Generally speaking, to rearrange the pages, long-press a page and drag it to a new spot. When you're done, tap the Back navigation icon.

>> The total number of Home screen pages is fixed. The maximum may be three, five, seven, or nine, depending on your phone.

>> On some phones, the far left Home screen page is the Google Now app.

>> Some phones allow you to set the primary Home screen page, which doesn't necessarily have to be the center Home screen page. I've seen different ways to accomplish this task. The most common one is to tap the Home icon in a thumbnail's preview, which is what's illustrated in Figure 20-2.

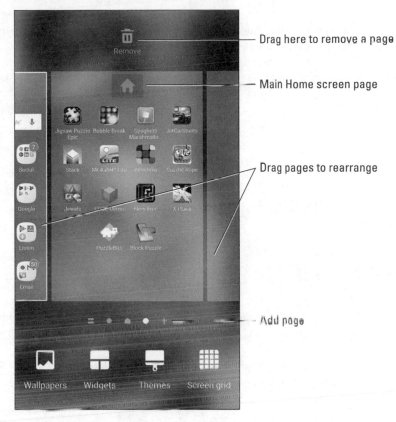

Drag here to remove a page

Main Home screen page

Drag pages to rearrange

Add page

FIGURE 20-2:
Manipulating
Home screen
pages.

Wallpapers   Widgets   Themes   Screen grid

Lock Screen Settings

The lock screen is different from the Home screen, although the two locations
share similar traits. As with the Home screen, you can customize the lock screen.
You can change the background, add app launcher shortcuts, and do all sorts of
tricks.

>> lock screen features might not be available for all Android phones.

>> For information on setting the screen lock, as well as controlling the lock
screen notifications, refer to Chapter 21.

Setting the lock screen wallpaper

For most Android phones, the lock screen wallpaper duplicates the Home screen
wallpaper. A few phones, however, let you set separate lock screen wallpaper.

To determine whether your phone provides for separate lock screen wallpaper, follow the same steps as outlined in the earlier section "Changing wallpaper." If you see a prompt or action bar to set the lock screen wallpaper, choose it. Then select for the lock screen a wallpaper that's different from the one on the Home screen.

>> Look for a prompt on the Set Wallpaper screen to see whether you can set the lock screen wallpaper or both lock screen and Home screen wallpaper at the same time.

>> Some phones feature a lock screen item in the Settings app. Choose that item to set the lock screen wallpaper. If not, try the Display item in the Settings app, and then choose Wallpaper.

Display Settings

The Display item in the Settings app deals with touchscreen settings. Two popular items worthy of your attention are the Brightness and Screen Timeout options.

The Display item might be titled Display and Wallpaper on your Samsung phone. It might be found under the Device tab.

Adjusting display brightness

The phone's touchscreen can be too bright, too dim, or just right. Which setting is best? That's up to you. Follow these steps:

1. **Open the Settings app and tap the Display item.**

   The item might be titled Display and Wallpaper.

2. **Choose Brightness Level.**

   This item might be titled Brightness, or a slider may be immediately visible.

3. **Adjust the slider to set the touchscreen's intensity.**

To have the phone adjust the brightness for you, locate the Adaptive Brightness item and set its master control to the On position. If you don't see an Adaptive Brightness item, look for an Auto check box.

TIP

You might also find a Brightness setting in the phone's quick settings. See Chapter 3 for information on accessing the quick settings.

Controlling the screen lock timeout

Your Android phone automatically locks itself after a given period of inactivity. For example, after 1 minute of your not touching the phone, the screen dims and then the phone locks. Follow these steps to control that duration:

1. **Open the Settings app and choose Display.**

2. **Choose the Sleep item.**

This item might be titled Screen Timeout.

3. **Set a timeout value from the list.**

The standard value is 1 minute.

The screen lock timeout measures inactivity. When you don't touch the screen or press a key, the timer starts ticking. A few seconds before the time-out value you set (refer to Step 3), the touchscreen dims. Then it locks. If you touch the screen before then, the sleep timer is reset.

Configuring the always-on touchscreen

Many phones feature an always-on or wake-up display: The touchscreen shows the current time and notifications either constantly on or they appear whenever the phone is moved. This convenience doesn't affect battery life, but the settings can be disabled, if you prefer. Obey these steps:

1. **In the Settings app, choose Display.**

2. **Set the master control by Ambient Display on or off.**

The On setting keeps the screen on or wakes it up; the Off option keeps the touchscreen dark while the phone is locked.

On some Samsung phones, choose Display and Wallpaper in Step 1, and then choose Always On Display. Adjust the master control, as described in Step 2.

Keyboard Settings

The standard Android keyboard is called the Google Keyboard. It offers some special features that supposedly make the onscreen typing experience more enjoyable. I'll leave it up to you to determine whether that's true.

Keyboard settings are located in the Settings app. Tap the Language & Input item. On some Samsung phones, that item is found in the Settings app on either the General tab or Controls tab.

Generating keyboard feedback

Typing seems to work better with an onscreen keyboard when the Haptic Feedback feature is active. To check this setting, follow these steps:

1. **Open the Settings app and choose Language & Input.**

2. **Choose Google Keyboard and then Preferences.**

    On Samsung phones, this item is titled Samsung Keyboard. It's the same thing. There is no need to choose Preferences for the Samsung keyboard.

3. **Use the master control to set Vibration on Keypress.**

    This item might be titled Vibration.

Activating predictive text

The Google Keyboard enables its predictive text automatically, but not every Android phone uses that keyboard. To ensure that the feature is active, follow these steps:

1. **Open the Settings app and choose Language & Input.**

2. **Choose Google Keyboard and then Text Correction.**

    Some Samsung phones show the Samsung Keyboard item. Choose it to view the Predictive Text setting.

3. **Ensure that the item Next-Word Suggestions or Predictive Text is enabled.**

    This item may instead be titled Show Suggestions.

If you can't locate a Next-Word Suggestions or Predictive Text item, it's most likely on all the time and can't be disabled.

See Chapter 4 for more information on using the Predictive Text feature.

Turning on gesture typing

Gesture typing allows you to create words by swiping your finger over the onscreen keyboard. Chapter 4 explains the details, although this feature may not be active on your phone. To ensure that it is, follow these steps:

1. **Open the Settings app and choose Language & Input.**

2. **Choose Google Keyboard, and then choose Gesture Typing.**

3. **Ensure that all Master Control icons are set to the On position.**

   Only the Enable Gesture Typing item needs to be enabled, though activating the other items enhances the experience.

For some Samsung phones, follow these steps:

1. **Open the Settings app and choose Language & Input.**

2. **Choose Samsung Keyboard.**

3. **Choose Keyboard Swipe.**

4. **Ensure that the option Continuous Input is chosen.**

Older Samsung phones use the item SwiftKey Flow in Step 3.

Activating dictation

Voice input should be enabled automatically on your Android phone, although that's no guarantee. The secret is to find the Dictation (Microphone) icon on the keyboard. The icon looks similar to the one shown in the margin.

TIP

On some Samsung keyboards, long-press the Multifunction key to locate the Dictation icon.

When you can find the Dictation key, and before you toss the phone into the garbage disposal, follow these steps to ensure that this feature is active:

1. **Open the Settings app and choose Language & Input.**

2. **Choose Google Voice Typing.**

3. **Ensure that the feature is enabled.**

   On many phones, the feature can't be disabled, so after choosing Google Voice Typing (refer to Step 2), you see a list of features.

Information on disabling the voice-typing word filter is found in Chapter 24.

Audio Adjustments

Yes, the phone makes noise. Incoming calls ring, you hear the sound of someone else on the line, and the various apps, alarms, and media beep, bleep, and blort. The Settings app is the place to go when the phone's sound needs fine-tuning.

Setting the volume

To adjust the volume for all potential sounds the phone can make, follow these steps:

1. **Open the Settings app and choose Sound & Notification.**

   This item might be titled Sound and Vibration or just Sound.

2. **On some Samsung phones, choose the Volume item.**

3. **Adjust the sliders to set the volume for various noises the phone makes.**

   These are the three common Android volume sliders:

   Media volume controls the sound for movies, videos, audio in the web browser, and so on.

   Alarm volume sets the intensity used for the Clock app's alarm.

   Ring Volume sets the phone's ringtone volume. This category includes incoming calls and notifications, although some phones may feature a separate slider for notifications.

   Other sliders may appear, such as System to adjust any volume not covered by the other categories.

4. **Slide the gizmo to the left to make a sound quieter; slide to the right to make a sound louder.**

   When you lift your finger, you hear a sound preview.

TIP

If you'd like the phone to vibrate on an incoming call, enable the Also Vibrate for Calls setting. This item might be titled Vibrate When Ringing, and it may be found on a separate Vibrations item on the Sound & Notification screen.

REMEMBER

Use the phone's Volume key to make the sound louder (up) or softer (down) for whatever you're listening to, such as music or an incoming call. The Volume key works even when the phone is locked.

Selecting the phone's ringtone

To set a new ringtone for your phone, or to simply confirm which ringtone it's using, follow these steps:

1. **Open the Settings app and choose Sound & Notification.**

   The item may be titled Sounds and Vibration.

2. **Choose Phone Ringtone or Ringtone.**

3. **Choose a ringtone from the list.**

   You hear the ringtone's preview.

4. **Tap OK to set the new ringtone, or tap CANCEL or the Back navigation icon to keep the current ringtone.**

The phone's notification ringtone is set separately from the incoming-call ringtone. To set the notification ringtone, choose the item Default Notification Ringtone or Notification Sound in Step 2.

TIP

>> If a ringtone app is installed, you may see a Complete Action Using card after Step 2. Choose the ringtone app to use it as a source, or choose Media Storage or Android System to choose one of the phone's preset ringtones. Refer to Chapter 19 for more details on default apps.

>> Various apps may set their own ringtones, such as text messaging ringtones and alert sounds for Facebook. These ringtones are set within the given app: Look for a Settings action in the app, either found on the navigation drawer or accessed by tapping the Action Overflow icon.

>> To disable the ringtone, choose None in Step 3. If you do so, I recommend activating the Vibration option so that the phone vibrates on an incoming call. Further, keep in mind that it's possible to temporarily disable sound on your phone. See Chapter 3.

>> Vibration settings are also found on the Sound & Notification screen in the Settings app.

Setting a contact's ringtone

Ringtones can be assigned by contact so that whenever your annoying friend Larry calls, you can have your phone yelp like a whiny puppy. Here's how to set a ringtone for a contact:

1. **Open the Contacts app.**

2. **Choose the contact for whom you want to assign a ringtone.**

3. **Edit the contact.**

   Tap the Edit icon (shown in the margin) or tap the EDIT button.

4. **Tap the Action Overflow and choose Set Ringtone.**

   On some phones, tap the More Fields item to view the Ringtone option.

5. **Select a ringtone from the list.**

   The ringtone you choose plays a brief preview.

6. **Tap OK to set the ringtone.**

   Some phones may automatically assign the ringtone.

7. **Save the edits.**

   Tap the Done icon or SAVE button to lock in your changes.

Whenever the contact calls, the phone rings using the ringtone you've specified.

To remove a specific ringtone for a contact, repeat the steps in this section, but in Step 5 choose Default Ringtone. (It's found at the top of the list.) This choice sets the contact's ringtone to be the same as the phone's ringtone.

Also see Chapter 6 for information on banishing a contact directly to voicemail.

Chapter 21

Android Phone Security

As more and more of your life is surrendered to the digital realm, the topic of security grows in importance. This concern extends directly to your phone, which is often home to your email, social networking, and other online accounts — including, potentially, important files and financial information. Don't take phone security lightly.

Lock Your Phone

The first line of defense for your Android phone is the screen lock. It can be simple, it can be complex, or it can be nonexistent. The choice is up to you.

Finding the screen locks

All screen locks on your phone are found in the Settings app, on the Choose Screen Lock screen. Heed these steps to visit that screen:

1. **Open the Settings app.**

2. **Choose Security.**

   This item may have another name, such as Lock Screen and Security. If you see both Security and Lock Screen items, choose Lock Screen.

   On some Samsung phones, you choose the Lock Screen item on the Device tab in the Settings app.

3. **Choose Screen Lock.**

   The item might also be titled Screen Lock Type, Set Up Screen Lock, or Change Screen Lock.

4. **Work any existing secure screen lock to continue.**

   Eventually, you see the Choose Screen Lock screen, which might instead be called Select Screen Lock.

The Choose Screen Lock screen lists several types of screen locks. Some are unique to your phone, and others are common Android screen locks, which include:

Swipe: Unlock the phone by swiping your finger across the touchscreen. This item might also be titled Slide.

Pattern: Trace a pattern on the touchscreen to unlock the phone.

PIN: Unlock the phone by typing a personal identification number (PIN).

Password: Type a password to unlock the phone.

Other screen locks may be available, including Face Unlock, Fingerprint, Signature, and even None, which isn't a lock at all.

REMEMBER

» The most secure lock types are the PIN and Password. Either screen lock type is required if the phone has multiple users, has a kid's account, or accesses a secure email server.

» The screen lock doesn't appear when you answer an incoming phone call. You're prompted, however, to unlock the phone if you want to use its features while you're on a call.

>> Your phone may prompt you to set the screen lock as the phone's power-on lock. That way, the lock appears when the phone is first turned on, as well as when you unlock the screen.

>> If you're in a panic, you can tap the EMERGENCY CALL button on the phone's Lock screen to bypass the screen lock and dial 911 or another emergency number.

WARNING

>> I know of no recovery method available if you forget your phone's PIN or password screen locks. If you use either one, write it down in an inconspicuous spot, just in case.

Removing a screen lock

You don't remove the screen lock on your Android phone. Instead, you replace it. Specifically, to remove the Pattern, PIN, and Password screen locks, set the Swipe lock. Follow the directions in the preceding section to get to the Choose Screen Lock screen.

>> You may be prompted for confirmation if you're opting to reset a secure screen lock to one that's less secure.

>> The phone prohibits you from removing a secure screen lock if the device is encrypted or accesses secure email or when other security features are enabled.

Setting a PIN

The PIN lock is second only to the Password lock as the most secure for your Android phone. To access the phone, you must type a PIN, or personal identification number. This type of screen lock is also employed as a backup for the less-secure screen unlocking methods, such as the Pattern lock.

The *PIN lock* is a code between 4 and 16 numbers long. It contains only digits, 0 through 9. To set the PIN lock for your Android phone, follow the directions in the earlier section "Finding the screen locks" to reach the Choose Screen Lock screen. Select PIN from the list of locks.

Use the onscreen keypad to type your PIN once, and then tap the CONTINUE button. Type the same PIN again to confirm that you know it. Tap OK. The next time you unlock the phone, you need to type the PIN to gain access.

Applying a password

The most secure way to lock the phone's screen is to apply a full-on password. Unlike a PIN, a password contains a combination of numbers, symbols, and uppercase and lowercase letters.

To set the password, choose Password from the Choose Screen Lock screen; refer to the earlier section "Finding the screen locks" for information on getting to that screen. The password you create must be at least four characters long. Longer passwords are more secure.

You're prompted to type the password whenever you unlock your phone or whenever you try to change the screen lock. Tap the OK button to accept the password you've typed.

Creating an unlock pattern

Perhaps the most popular, and certainly the most unconventional, screen lock is the Pattern lock. You must trace a pattern on the touchscreen to unlock the phone. To create a Pattern lock, follow these steps:

1. **Summon the Choose Screen Lock screen.**

    Refer to the earlier section "Finding the screen locks."

2. **Choose Pattern.**

    If you haven't yet set a pattern, you may see the tutorial describing the process; tap the Next button to skip merrily through the dreary directions.

3. **If the phone prompts you for Secure Start-Up, choose the option Require Pattern to Start Device, and then tap the CONTINUE button.**

    Alternatively, you can choose No Thanks, which is less secure.

4. **Trace an unlock pattern.**

    Use Figure 21-1 as your guide. You can trace over the dots in any order, but you can trace over a dot only once. The pattern must cover at least four dots.

5. **Tap the CONTINUE button.**

6. **Redraw the pattern.**

7. **Tap the CONFIRM button.**

    You may be required to type a PIN or password as a backup to the Pattern lock. If so, follow the onscreen directions to set that lock as well.

TIP

To ensure that the pattern appears on the Lock screen, place a check mark by the Make Pattern Visible option. For even more security, you can disable this option, but you must remember how — and where — the pattern goes.

Also: Clean the touchscreen! Smudge marks can betray your pattern.

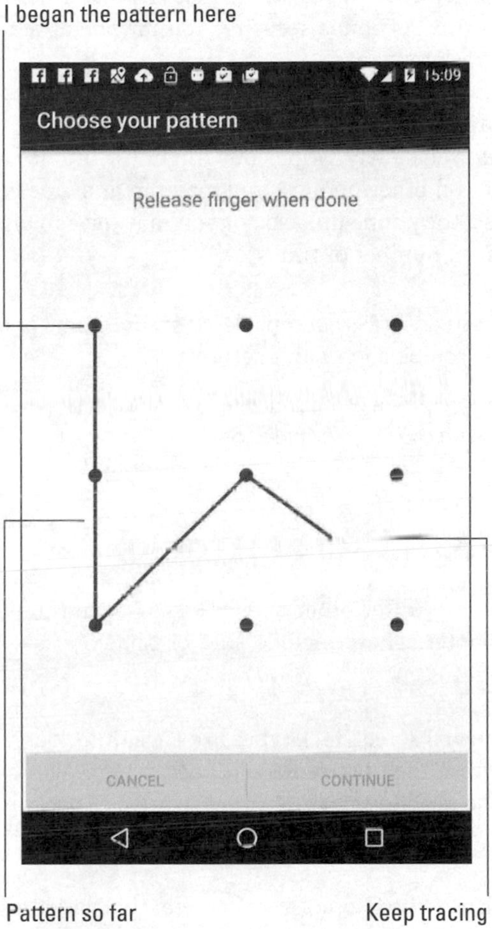

I began the pattern here

Choose your pattern

Release finger when done

CANCEL   CONTINUE

FIGURE 21-1:
Set the unlock
pattern.

Pattern so far   Keep tracing

Using a fingerprint lock

Some phones come with fingerprint scanners. The physical Home button on a Samsung phone may double as a fingerprint reader. Other phones may feature a fingerprint reader on the back.

The Fingerprint screen lock is a unique type of screen lock, so configuring this feature works differently on each phone. You may find the controls with the other screen locks on the Choose Screen Lock screen, or you may find a Fingerprint item on the main Security screen: Open the Settings app and choose Security.

After you select the Fingerprint screen lock, obey the directions on the screen. You may have to swipe or tap your finger on the reader a few times before it accurately records your fingerprint. As bonus security, you may be asked to supply a PIN or Password lock as a backup.

To use the Fingerprint lock, swipe or tap your finger on the fingerprint reader. On Samsung phones, you slowly swipe your finger (or thumb) over or tap directly on the Home button. On other phones, you may tap or swipe your finger over the reader. Upon success, the phone unlocks, or you may have to type the PIN or password if you fail a given number of times.

>> The Fingerprint screen lock is *not* considered secure. Some phones may let you use it only in combination with a secure lock.

>> You can answer the phone without having to unlock it, so just swipe to answer and don't mess with the fingerprint reader.

Setting unusual screen locks

Your Android phone may offer other screen locks beyond the conventional locks described in this chapter. These include silly or fancy locks — perhaps not that secure, but fun and different.

Among the more unusual screen lock types are Face Unlock and Signature. Choose these screen locks from the Choose Screen Lock screen. Work through the setup process. You may also need to set a PIN or password as a backup to the less-secure screen lock types.

>> The Face Unlock uses the phone's front camera. To unlock the device, you stare at the screen. So long as you haven't had any recent, major plastic surgery, the phone unlocks.

>> The Signature lock is unique to the Samsung Galaxy Note phone. Use the S Pen to scribble your John Hancock on the touchscreen. The phone unlocks.

Other Phone Security

Beyond locking the screen, other tools are available to help you thwart the Bad Guys and keep safe the information in your phone. Beyond that, you can even employ methods to locate a lost or stolen phone.

Controlling lock screen notifications

Your phone may allow for notifications to be displayed on the lock screen, similar to how notifications are listed on the notifications drawer. You can set which notifications appear and how they're displayed, which helps add to the phone's security.

To configure lock screen notifications, follow these steps:

1. **Open the Settings app.**

2. **Choose Sound & Notification.**

   This item might be titled Sounds and Notifications.

3. **Choose When Device Is Locked.**

   Another title for this setting is Notifications on lock Screen

4. **Select a lock screen notification level.**

   Up to three settings are available:

   - Show All Notification Content
   - Hide Sensitive Notification Content
   - Don't Show Notifications at All

   The names of these settings may be subtly different on your phone.

5. **Choose a notification level.**

On some Samsung phones, choose Lock Screen and Security in Step 2, and then choose Notifications on the Lock Screen. Choose the item Content on Lock Screen to view three options, similar to those shown in Step 4.

>> The Hide Sensitive Notification Content option (refer to Step 4) appears only when a secure screen lock is chosen.

>> Double-tap a lock screen notification to open its related app and view more details. You must unlock the screen first, and then the app opens.

Adding owner info text

Suppose that you lose your Android phone. Wouldn't it be nice if a good Samaritan found it? What would be even more helpful is information on the Lock screen to help that kind person find you and return your gizmo. That information is called the *owner info text.*

To add owner info text to your Android phone's Lock screen, follow these steps:

1. **Open the Settings app.**

2. **Choose the Security or Lock Screen category.**

   On some Samsung phones, the Lock Screen category is found on the Device tab.

3. **Choose Lock Screen Message.**

   This item might also be titled Owner Info or Owner Information.

   On some Samsung phones, choose Info and App Shortcuts, and then choose Owner Information.

4. **Type text in the box.**

   You can type more than one line of text, though the information is displayed on the Lock screen as a single line.

   TIP

5. **Tap the SAVE or DONE button.**

Whatever text you type in the box appears on the Lock screen. Therefore, I recommend typing something useful: your name, address, another phone number, an email address, or similar vital information. This way, should you lose your phone and an honest person find it, he can get it back to you.

The owner info may not show up when None is selected as a screen lock.

Finding a lost phone

Someday, you may lose your beloved Android phone. It might be for a few panic-filled seconds, or it might be forever. The hardware solution is to weld a heavy object to the phone, such as an anvil. Alas, that strategy kind of defeats the entire mobile/wireless paradigm.

To quickly locate your Android phone, follow these steps while using a computer:

1. **Open a web browser, such as Google Chrome.**

2. **Visit the main Google search page:** www.google.com

3. **Type** find my phone **and press the Enter key.**

4. **If prompted, sign in to your Google (Gmail) account.**

   Your phone's location appears on the screen.

Another solution is to use a cell phone locator service.

Cell phone locator services employ apps that use the phone's mobile data signal as well as its GPS to help locate the missing gizmo. These types of apps are available at Google Play. One that I've tried and recommend is Lookout Mobile Security.

Lookout features two different apps. One is free, which you can try, to see whether you like it. The paid app offers more features and better locating services. As with similar apps, you must register at a website to help you locate your phone, should it go wandering.

Encrypting the phone

When the information on your beloved phone must be really, really secure, you can take the drastic step of encrypting its storage. Before considering this option, however, be aware that it's not currently possible to remove encryption. After encryption is applied, it's stuck forever like a regrettable, drunken tattoo.

Start the encryption process by applying a secure screen lock, such as the PIN or password. Refer to the first part of this chapter for details.

Second, ensure that the phone is either plugged in or fully charged. Encryption takes a while — up to several hours if your phone's storage is rather full — and you don't want your gizmo pooping out before the process is complete.

Third, follow these steps to encrypt the phone's internal storage:

1. **Open the Settings app and choose Security.**

   On some Samsung phones, choose Lock Screen and Security.

2. **Choose Encrypt Phone.**

   This item may be titled Protect Encrypted Data.

3. **Tap the ENCRYPT PHONE button.**

   If you don't see this button, or you can't get to the proper screen, the phone's data is already encrypted.

4. **Wait.**

After it's encrypted, only by unlocking the phone can anyone access its storage. Further, you need to work the secure screen lock every time you turn on the phone.

Performing a factory data reset

The most secure thing you can do with the information on your Android phone is to erase it all. The procedure, known as a factory data reset, effectively restores the device to its original state, fresh out of the box.

WARNING

A factory data reset is a drastic event. It not only removes all information from storage but also erases all your accounts. Don't take this step lightly! In fact, if you're using this procedure to cure an ill, I recommend first getting support. See Chapter 23.

When you're ready to erase all data from the phone, follow these steps:

1. **Start the Settings app and choose Backup & Reset.**

On Samsung phones, tap the General tab to locate the Backup and Reset item.

2. **Choose Factory Data Reset.**

3. **Tap the RESET PHONE button.**

4. **If prompted, work the screen lock.**

This level of security prevents others from idly messing with your beloved phone.

5. **Tap the ERASE EVERYTHING or DELETE ALL button to confirm.**

All the information you've set or stored on the phone is purged, including all your accounts, any apps you've downloaded, music — everything.

TIP

Practical instances when this action is necessary include selling your phone, giving it to someone else to use, and exchanging it for a new phone.

Chapter 22

On the Road Again

You're in a land far, far away. The sun shines warmly upon your face. A gentle breeze wafts over crashing waves. You wiggle your toes in the soft, grainy sand. And the number-one thought on your mind is, "Can my Android phone get a signal?"

The vernacular term is *bars*, as in the signal bars you see when the phone is actively communicating with the mobile data network. That network pretty much covers the globe these days. That means from Cancun to the Gobi Desert, you'll probably be able to use your phone. How that happens and how to make it happen less expensively are things traveling Android phone users must know.

Where the Phone Roams

The word *roam* takes on an entirely new meaning when applied to a cell phone. It means that your phone receives a cell signal whenever you're outside the phone carrier's operating area. In that case, your phone is roaming.

Roaming sounds handy, but there's a catch: It almost always involves a surcharge for using another cellular service — an unpleasant surcharge.

R.ıl| Relax: Your Android phone alerts you whenever it's roaming. The Roaming icon appears at the top of the screen, in the status area, whenever you're outside your cellular provider's signal area. The icon differs from phone to phone, but generally the letter *R* figures in it somewhere, similar to what's shown in the margin.

There's little you can do to avoid roaming surcharges when making or receiving phone calls. Well, yes: You can wait until you're back in an area serviced by your primary cellular provider. You can, however, altogether avoid using the other mobile data network while roaming. Follow these steps:

1. **Open the Settings app.**

2. **In the Wireless & Networks area, tap the More item.**

3. **Choose Cellular Networks.**

    If you see a Mobile Networks item in the Wireless and Networks area, choose it instead and then choose Global Data Roaming Access.

4. **Ensure that the Data Roaming setting is disabled or denied.**

On some Samsung phones, follow these steps:

1. **Open the Settings app and choose Mobile Networks.**

2. **Choose Data Roaming Access.**

3. **Select the option Deny Data Roaming Access and tap OK.**

Your phone can still access the Internet over the Wi-Fi connection when it roams. Setting up a Wi-Fi connection doesn't affect the mobile data network connection, because the phone prefers to use Wi-Fi. See Chapter 17 for more information about Wi-Fi.

Another network service you might want to disable while roaming has to do with multimedia, or MMS, text messages. To avoid surcharges from another cellular network for downloading an MMS message, follow these steps:

1. **Open the phone's text messaging app.**

2. **Ensure that you're viewing the main screen, not an individual message thread.**

3. **Tap the Action Overflow icon or the MORE button.**

4. **Choose Settings.**

5. **Choose Advanced or More Settings.**

On some Samsung phones, you must also choose the Multimedia Messages item.

6. **Ensure that the Roaming Auto Retrieve option is disabled.**

The lock screen may also announce that the phone is roaming. You might see the name of the other cellular network displayed. The text *Emergency Calls Only* might also appear.

TIP

If you're greatly concerned about roaming while overseas, place the phone into Airplane mode, as discussed elsewhere in this chapter.

International Calling

A phone is a bell that anyone in the world can ring. To prove it, all you need is the phone number of anyone in the world. Use your phone to dial that number and, as long as you both speak the same language, you're talking!

To make an international call with your Android phone, you merely need to know the foreign phone number. The number includes the international country-code prefix, followed by the number. For example:

01-234-56-789

Before dialing the international country-code prefix (01 in this example), you must first type a plus (+). The + symbol is the country exit code, which must be dialed to flee the national phone system and access the international phone system. For example, to dial Finland on your phone, type +358 and then the number in Finland. The +358 is the exit code (+) plus the international code for Finland (358).

To type the + character, press and hold down the 0 key on the Phone app's dialpad. Then type the country prefix and the phone number. Tap the Dial icon to place the call.

>> Dialing internationally involves surcharges, unless your cell plan provides for international calling.

>> International calls fail for a number of reasons. One of the most common is that the recipient's phone service blocks incoming international calls.

>> Another reason that international calls fail is the zero reason: Oftentimes, you must leave out any zero in the phone number that follows the country code. So, if the country code is 254 for Kenya and the phone number starts with 012,

you dial +254 for Kenya and then 12 and the rest of the number. Omit the leading zero.

>> Know which type of phone you're calling internationally — cell phone or landline. The reason is that an international call to a cell phone might involve a surcharge that doesn't apply to a landline.

WARNING

>> The + character isn't a number separator. When you see an international number listed as 011+20+xxxxxxx, do not insert the + character in the number. Instead, type +20 and then the rest of the international phone number.

>> Most cellular providers add a surcharge when sending a text message abroad. Contact your cellular provider to confirm the text message rates. Generally, you find two rates: one for sending and another for receiving text messages.

TIP

>> If texting charges vex you, remember that email has no associated per-message charge. You also have alternative ways to chat, such as Google Hangouts and Skype; Skype can also be used to place cheap international calls. See Chapter 11.

REMEMBER

>> In most cases, dialing an international number involves a time zone difference. Before you dial, be aware of what time it is in the country or location you're calling. The Clock app can handle that job for you: Summon a clock for the location you're calling and place it on the Clock app's screen.

You Can Take It with You

You can take your Android phone with you anywhere you like. How it functions may change depending on your environment, and you can do a few things to prepare before you go. Add these items to your other travel checklists, such as taking cash, bringing an ID, and preparing to wait in inspection lines.

Preparing to leave

Unless you're being unexpectedly abducted, you should prepare several things before leaving on a trip with your phone.

Most important, of course, is to charge the thing. You probably charge your phone overnight anyway, but if you're leaving in the middle of the day, plug in your phone to ensure that the charge will last the journey.

Also, consider loading up on some reading material, music, and a few new apps before you go. Nothing beats getting eBooks for the road. Visit Google Play to shop for movies, music, and books before you go. Perhaps even load up a new game that you can learn to play.

REMEMBER

TIP

>> If you plan to read books, listen to music, or watch a video while on the road, consider downloading that media to your phone before you leave. See Chapter 16 for information.

>> Most major airlines offer travel apps. These apps generate notifications for your schedule and provide timely gate changes or flight delays; plus, you can use the phone as your e-ticket. Search Google Play to see whether your preferred airline offers an app.

Arriving at the airport

I'm not a frequent flier, but I am a nerd. The most amount of junk I've carried with me on a flight is two laptop computers and three cell phones. I know that's not a record, but it's enough to warrant the following list of travel tips, all of which apply to taking your phone with you on an extended journey:

>> Bring the phone's AC adapter and USB cable with you. Put them in your carry-on luggage.

>> Nearly all major airports feature USB chargers, so you can charge the phone in an airport, if you need to. Even though you need only the cable to charge, bring along the AC adapter anyway.

>> At the security checkpoint, place your phone in a bin by itself or with other electronics (tablets, laptops). Even if you've been preapproved, you must remove the phone from your pocket.

>> Open the phone's web browser to use the airport's Wi-Fi service. Most airports don't charge for the service, although you may have to agree to terms by visiting the airport's website and watching an advertisement or agreeing to the service terms.

Flying with your phone

After the flight crew closes the airplane's door, you must turn off your phone or place it into Airplane mode. You can still use the phone, just not to make phone calls, send text messages, or check the map.

Follow these steps to place your phone into Airplane mode:

1. Open the Settings app.

2. Choose the More item in the Wireless & Networks section.

If you find an Airplane Mode item in the Wireless and Networks section, choose that item instead.

3. Enable Airplane mode.

Slide the master control to the On position, or place a check mark by the Airplane Mode item.

While the phone is in Airplane mode, a special icon appears in the status area, similar to the one shown in the margin. You might also see the text *No Service* on the phone's lock screen.

TIP

>> You can quickly enter Airplane mode by choosing its quick setting: Swipe down the notifications drawer and choose Airplane Mode. See Chapter 3 for more details on accessing your phone's quick settings.

>> Some phones feature an Airplane Mode action on the Device Options menu: Press and hold the Power/Lock key as though you're turning off the phone. Choose Airplane Mode.

>> Some Android phones feature Sleep mode, in which case you can place your phone into Sleep mode for the duration of a flight. The phone wakes up faster from Sleep mode than it does when you turn it on.

REMEMBER

>> You can still use Wi-Fi while aloft, as long as that service is available and you're willing to pay for it. You must activate Wi-Fi after enabling Airplane mode. See Chapter 17 for information on turning on the phone's Wi-Fi radio.

TECHNICAL
STUFF

>> Airplane mode disables all of the phone's wireless radios: the cellular radio, Wi-Fi, Bluetooth, NFC, and GPS. You can, however, reenable Wi-Fi, although using the other radios is forbidden while flying.

Getting to your destination

After you arrive at your destination, the phone may update the date and time according to your new location. One additional step you may want to take is to set the phone's time zone. By doing so, you ensure that your schedule adapts properly to your new location.

To change the phone's time zone, follow these steps:

1. Open the Settings app.

2. Choose Date and Time.

On some Samsung phones, tap the General tab to locate the Date and Time item.

3. Ensure that the Automatic Time Zone setting is enabled.

If the phone's time isn't updated, continue by choosing the Select Time Zone item. Pluck the current time zone from the list.

If you've set appointments for your new location, visit the Calendar app to ensure that their start and end times have been properly adjusted. If you're prompted to update appointment times based on the new zone, do so.

REMEMBER

When you're done traveling, or when you change time zones again, ensure that the phone is updated as well. When the Automatic Time Zone setting isn't available, follow the steps in this section to reset the phone's time zone.

The Android Phone Goes Abroad

Yes, your Android phone works overseas. The world uses similar types of cellular networks, so the odds are good that your phone will work immediately in whichever county you find yourself. Your concerns should be overseas roaming, charging the phone in a foreign country, and how to get on the Internet.

Using your Android phone overseas

The easiest way to use a cell phone abroad is to rent or buy one in the country where you plan to stay. I'm serious: Often, international roaming charges are so high that it's cheaper to simply buy a temporary cell phone wherever you go, especially if you plan to stay there for a while.

When you opt to use your own phone rather than buy a local phone, things should run smoothly — if a compatible cellular service is in your location. Not every Android phone uses the same network and, of course, not every foreign country uses the same cellular network. Things must match before the phone can work. Plus, you may have to deal with foreign carrier roaming charges.

The key to determining whether your phone is usable in a foreign country is to turn it on. The name of that country's compatible cellular service should show up on the phone's lock screen. So where your phone once said *Verizon Wireless*, it may say *Wambooli Telcom* when you're overseas.

>> You receive calls on your cell phone internationally as long as the phone can access the network. Your friends need only dial your cell phone number as they normally do; the phone system automatically forwards your calls to wherever you are in the world.

>> The person calling you pays nothing extra when you're off romping the globe with your Android phone. Nope — you pay extra for the call.

>> While you're abroad, you need to dial internationally. When calling home (for example, the United States), you need to use a 10-digit number (phone number plus area code). You may also be required to type the country exit code when you dial. See the earlier section "International Calling."

>> When in doubt, contact your cellular provider for tips and other information specific to whatever country you're visiting.

>> Be sure to inquire about texting and cellular data (Internet) rates while you're abroad.

>> Also see the section "Where the Phone Roams," earlier in this chapter. Roaming charges definitely apply overseas.

Charging the phone in another country

You can easily attach a foreign AC power adapter to your phone's AC power plug. You don't need a voltage converter; just an adapter. After the adapter is attached, plug your phone into those weirdo overseas power sockets without the risk of blowing up anything.

Accessing Wi-Fi in foreign lands

Using an Android phone over a Wi-Fi network abroad incurs no extra fees (unless data roaming is on, as discussed earlier in this chapter). The same protocols and standards are used everywhere, so if the phone can access Wi-Fi at your local Starbucks, it can access Wi-Fi at the Malted Yak Blood Café in Wamboolistan. As long as Wi-Fi is available, your Android phone can use it.

>> Internet cafés are more popular overseas than in the United States. They are the best locations for connecting your phone and catching up on life back home.

>> Many overseas hotels offer free Wi-Fi service, although the signal may not reach into every room. Don't be surprised if you can use the Wi-Fi network only while you're in the lobby.

>> Use your Android phone to make phone calls overseas by getting some Skype Credit. Skype's international rates are quite reasonable. The calls are made over the Internet, so when the phone has Wi-Fi access, you're good to go. See Chapter 11 for more information on making Skype calls.

Chapter 23

Maintenance, Troubleshooting, and Help

Maintenance is that thing you were supposed to remember to do but you didn't do, and that's why you need help and troubleshooting advice. Don't blame yourself; no one likes to do maintenance. Okay, well, I like maintaining my stuff. I even change the belt on my vacuum cleaner every six months. Did you know that the vacuum cleaner manual tells you to do so? Probably not. I read that in *Vacuum Cleaners For Dummies*. This book is *Android Phones For Dummies*, which is why it contains topics on maintenance, troubleshooting, and help for Android phones, which lack belts that you should change every six months.

Regular Phone Maintenance

Relax. Unlike draining the lawnmower's oil once a year, regular maintenance of an Android phone doesn't require a drip pan or a permit from the EPA. In fact, an Android phone requires only two basic regular maintenance tasks: cleaning and backing up.

Keeping it clean

You probably already keep your phone clean. I use my sleeve to wipe the touchscreen at least a dozen times a day. Of course, better than your sleeve is something called a microfiber cloth. This item can be found at any computer or office-supply store.

WARNING

>> Never use ammonia or alcohol to clean the touchscreen. These substances damage the phone. If you must use a cleaning solution, select one specifically designed for touchscreens.

>> If the screen continually gets dirty, consider adding a screen protector. This specially designed cover prevents the screen from getting scratched or dirty while still allowing you to interact with the touchscreen. Ensure that the screen protector you obtain is intended for use with your specific phone.

>> You can also find customized cell phone cases, belt clips, and protectors, which can help keep the phone looking spiffy. Be aware that these items are mostly for decorative or fashion purposes and don't even prevent serious damage when you drop the phone.

Backing up your phone

For a majority of the information on your phone, backup is automatic. Your Google account takes care of the Gmail, calendar, contacts — even your apps, music, eBooks, and movies. Most of the stuff on your phone is backed up automatically.

To confirm that your phone's information is being backed up, heed these steps:

1. Open the Settings app and choose Accounts.

On some phones, the accounts are listed directly on the main Settings app screen.

2. Choose Google to access your Google account.

3. **Ensure that the master control by each item in the list is enabled.**

These are the items that synchronize between the phone and your Google account on the Internet.

You're not done yet!

4. **Tap the Back navigation until you're returned to the main Settings app screen.**

5. **Choose Backup & Reset.**

6. **Ensure that the item Back Up My Data is enabled.**

Beyond your Google account, which is automatically backed up, the rest of the information can be manually backed up. You can copy files from the phone's internal storage to the cloud or your computer as a form of backup. See Chapter 18 for information on coordinating files between your phone and a computer.

Updating the system

Every so often, a new version of your phone's operating system is released. It's an Android update because Android is the name of the phone's operating system, not because your phone thinks that it's a type of robot.

Whenever an update occurs, you see a notification message indicating that a system upgrade is available. My advice: Install the update and get it over with. Don't dally.

TIP

To manually check for updates, follow these steps:

1. **Open the Settings app.**

2. **Choose About Phone and then choose System Updates.**

On Samsung phones, the item is titled Download Updates Manually.

If any updates are pending, you see them listed. Tap the RESTART & INSTALL button. Otherwise, you can tap the CHECK FOR UPDATE button. The button isn't magic and won't force an update if one isn't available.

>> If possible, connect the phone to a power source during a software update. You don't want the battery to die in the middle of the operation.

>> Non-Android system updates might also be issued. For example, the phone's manufacturer may send out an update to the device's guts. This type of update is often called a *firmware* update. As with Android updates, my advice is to accept all firmware updates.

Battery Care and Feeding

Perhaps the most important item you can monitor and maintain on your Android phone is its battery. The battery supplies the necessary electrical juice by which the phone operates. Without battery power, your phone is about as useful as a tin-can-and-a-string for communications. Keep an eye on the battery.

Monitoring the battery

Your phone displays its current battery status at the top of the screen, in the status area, next to the time. The icons used are similar to those shown in Figure 23-1.

Battery is fully charged and happy.

Battery is being used but starting to drain.

Battery getting low; you should charge!

Battery frighteningly low; stop using and charge at once!

FIGURE 23-1:
Battery status icons.

Battery is charging.

You might also see the icon for a dead or missing battery, but for some reason I can't get my phone to turn on and display that icon.

>> Heed those low-battery warnings! The phone generates a notification whenever the battery power gets low. The phone generates another notification whenever the battery gets very low.

>> When the battery is too low, the phone shuts itself off.

TIP

>> The best way to deal with a low battery is to connect the phone to a power source. Either plug the phone into a wall socket, or use the USB cable to connect the phone to a computer. The phone charges itself immediately; plus, you can use the phone while it's charging.

>> The phone charges more efficiently when it's plugged into a wall socket rather than a computer.

>> You don't have to fully charge the phone to use it. If you have 20 minutes to charge and the power level returns to only 70 percent, that's great. Well, it's not great, but it's far better than a 20 percent battery level.

TECHNICAL STUFF

>> Battery percentage values are best-guess estimates. Just because you talked for two hours and the battery shows 50 percent doesn't mean that you're guaranteed two more hours of talking. Odds are good that you have much less than two hours. In fact, as the percentage value gets low, the battery appears to drain faster.

Determining what is drawing power

The Battery screen informs you of the phone's battery usage over time, as well as which apps have been consuming power. Figure 23-2 illustrates such a screen.

To view the Battery screen on your phone, open the Settings app and choose the Battery item. On some Samsung phones, tap the General tab in the Settings app to locate the Battery item.

Tap an item in the list to view its details. For some items, the details screen lets you adjust power settings.

The number and variety of items listed on the battery use screen depend on what you've been doing with your phone between charges and how many different apps you've been using. Don't be surprised if an item (such as the Play Books app) doesn't show up in the list. Not every app uses a lot of battery power.

Extending battery life

Here's a smattering of things you can do to help prolong battery life for your Android phone:

Dim the screen. The display is capable of drawing down quite a lot of battery power. Although a dim screen can be more difficult to see, especially outdoors, it definitely saves on battery life. See Chapter 20 for information on adjusting the screen brightness.

Current battery charge and state

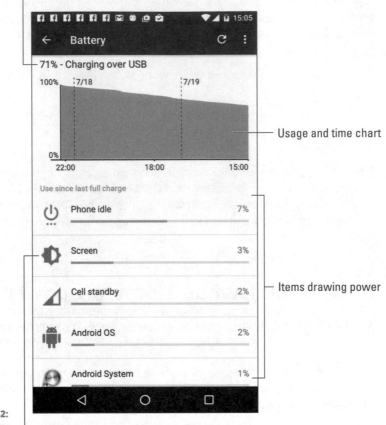

— Usage and time chart

— Items drawing power

FIGURE 23-2:
Battery
usage info.

Tap to view usage and change settings

Disable vibration options. The phone's vibration is caused by a teensy motor. Although you don't see much battery savings by disabling the vibration options, it's better than no savings. To turn off vibration, follow these steps:

1. **Open the Settings app and choose Sound & Notification.**

2. **Disable one or all of the vibration settings.**

Most phones lack a single setting for vibration. Instead, you find settings such as Vibrate While Ringing, Vibration Intensity, and Vibrate on Touch.

Deactivate Bluetooth. When you're not using Bluetooth, turn it off. See Chapter 17 for information on Bluetooth.

Manage battery performance. Many Android phones have battery-saving software built in. You can access the software from a special app or from the Battery item in the Settings app. Similar to power management on a computer, battery-performance management involves turning phone features on or off during certain times of the day. Third-party battery management apps are also available from Google Play. See Chapter 16.

Help and Troubleshooting

Getting help isn't as bad as it was in the old days. Back then, you could try two sources for technological help: the atrocious manual that came with your electronic device or a phone call to the guy who wrote the atrocious manual. It was unpleasant. Today, the situation is better. You have many resources for solving issues with your gizmos, including your Android phone.

Fixing random and annoying problems

Aren't all problems annoying? A welcome problem doesn't exist, unless the problem is welcome because it diverts attention from another, preexisting problem. And random problems? If problems were predictable, they would serve in office.

General trouble

For just about any problem or minor quirk, consider restarting the phone: Turn off the phone, and then turn it on again. This procedure will likely fix a majority of the annoying and quirky problems you encounter when using an Android phone.

TIP

Some Android phones feature the Restart action on the Phone Options menu: Press and hold down the Power/Lock key to see this menu. If a Restart command is present, use it to restart the phone and fix (you hope) whatever has gone awry.

Connection woes

Sometimes the data connection drops but the phone connection stays active. Check the status bar. If you see bars, you have a phone signal. When you don't see the 4G, H+, Wi-Fi, or other Internet-connection icon, the phone has no Internet access.

Do expect the mobile data signal to change. You may observe the status icon change from 4G LTE to 3G to even the dreaded 1X or — worse — nothing, depending on the strength and availability of the mobile data network. Wait, and it should come back around. If not, the mobile data network might be down, or you may simply be in an area with lousy service. Consider changing your location.

For Wi-Fi connections, ensure that Wi-Fi is set up properly and working. This process usually involves pestering the person who configured the Wi-Fi router or, in a coffee shop, bothering the cheerful person with the tattoos and piercings who serves you coffee.

Be aware that some Wi-Fi networks have a "lease time" after which the phone is disconnected. If so, follow the directions in Chapter 17 for turning off Wi-Fi and then turn it on again. That often solves the issue.

Another problem could be that the Wi-Fi router doesn't recognize your phone. It may be an older router that can't handle an abundance of wireless devices. If this situation describes your home or office network, and the Wi-Fi router is over five years old, consider getting a newer router.

Music is playing and you want it to stop

It's awesome that the phone continues to play music while you do other things. Getting the music to stop quickly, however, requires some skill. You can access the play controls for the Play Music app from a number of locations. They're found on the lock screen, for example. You can also find them on the notifications drawer.

An app has run amok

Sometimes, apps that misbehave let you know. You see a warning on the screen announcing the app's stubborn disposition. When that happens, tap the Force Quit button to shut down the app. Then say "Whew!"

To manually shut down an app, refer to Chapter 19.

You've reached your wit's end

When all else fails, you can do the drastic thing and perform a factory data reset on your phone. Before committing to this step, I recommend that you contact support as described in the next section.

Refer to Chapter 21 for details on the factory data reset.

Getting help and support

Never discount two important sources of support for your Android phone: your cellular provider and the phone's manufacturer. Between the two, I recommend contacting the cellular provider first, no matter what the problem.

The Help app

Some phones come supplied with the Help app. The app may be called Help or Help Center or something similar, and it may not be the kind of avuncular, well-written help you get from this book, but it's better than nothing.

>> A Guided Tour or Tutorial app may also be available, which helps you understand how to work some of the phone's interesting features.

>> Also look for a Help document or phone manual eBook in the Play Books app.

>> The Settings app features the Search icon, which helps you locate specific settings without knowing exactly under which category the item might be found.

Cellular support

Contact information for both the cellular provider and phone manufacturer is found in the material you received with the phone. In Chapter 1, I recommend that you save those random pieces of paper. You obviously have read that chapter and followed my advice, so you can easily find that information.

Okay, so you don't want to go find the box, or you didn't heed my admonition and you threw out the box. Table 23-1 lists contact information on U.S. cellular providers. The From Cell column lists the number you can call by using your Android phone; otherwise, you can use the toll-free number from any phone.

TABLE 23-1 **U.S. Cellular Providers**

Provider	From Cell	Toll-free	Website
AT&T	611	800-331-0500	www.att.com/esupport
Sprint Nextel	*2	800-211-4727	sprint.com
T-Mobile	611	877-453-1304	www.t-mobile.com/Contact.aspx
Verizon	611	800-922-0204	verizonwireless.com/support

TIP

Your cellular provider's Help phone number might already be found in the phone's address book app. Look for it there if it's not listed in the table.

Manufacturer support

The second source of support is the phone's manufacturer, such as Samsung or LG. Information about support can be found in those random papers and pamphlets included in the phone's box. If not, refer to Table 23-2 for contact information for various Android handset manufacturers.

TABLE 23-2

Android Phone Manufacturers

Manufacturer	Website
HTC	www.htc.com/us/support
LG	www.lg.com/us/support
Motorola	www.motorola.com
Samsung	samsung.com/us/mobile/cell-phones

App support

For app issues, contact the developer. Unlike finding developer support for computer software, accessing an app developer is cinchy. Follow these steps:

1. **Open the Play Store app.**

2. **Tap the Side Menu icon to display the navigation drawer.**

3. **Choose My Apps & Games.**

   If you don't see this item, choose Apps & Games and then repeat Steps 2 and 3.

4. **Tap the entry for the specific app, the one that's bothering you.**

5. **Choose the Send Email item.**

   Swipe the screen bottom-to-top to scroll down and locate the Send Email item.

REMEMBER

Contacting the developer is no guarantee that you'll get a response.

Google Play support

For issues with Google Play itself, contact support.google.com/googleplay

Android Phone Q&A

I love Q&A! Not only is it an effective way to express certain problems and solutions, but some of the questions might also cover topics I've been yearning to write about.

"The touchscreen doesn't work!"

A touchscreen requires a human finger for proper interaction. The phone interprets complicated and magical physics between the human finger and the phone to determine where the touchscreen is being touched. Bottom line: You can't wear gloves and use the touchscreen unless they are specific touchscreen gloves.

The touchscreen might also fail when the battery charge is low or when the phone has been physically damaged.

"The screen is too dark!"

Android phones feature a teensy light sensor on the front. If the phone's Adaptive Brightness feature is active, the sensor adjusts the touchscreen's brightness based on the amount of ambient light at your location. If the sensor is covered, the screen can get very, very dark.

Ensure that you aren't unintentionally blocking the light sensor. Avoid buying a case or screen protector that obscures the sensor.

Also see Chapter 20 for information on setting the screen brightness.

"The battery doesn't charge!"

Start from the source: Is the wall socket providing power? Is the cord plugged in? The cable may be damaged, so try another cable.

When charging from a USB port on a computer, ensure that the computer is turned on. Computers provide no USB power when they're turned off. The phone may charge only when connected to a powered USB hub, such as those found directly on the computer console.

"The phone gets so hot that it turns itself off!"

Yikes! An overheating phone can be a nasty problem. Judge how hot the phone is by seeing whether you can hold it in your hand: When the phone is too hot to hold, it's too hot. If you're using the phone to keep your coffee warm, the phone is too hot.

Turn off the phone and let it cool.

If the overheating problem continues, have the phone looked at for potential repair.

"The phone won't do Landscape mode!"

Not all apps can change their orientation, especially some games. Also, not every Android phone reorients the Home screen. One app that definitely does Landscape mode is the web browser, which is described in Chapter 10.

TIP

Android phones have a quick setting you can check to confirm that screen rotation is enabled or to lock the screen at a specific orientation. See Chapter 3 for information on quick settings.

5

The Part of Tens

Chapter 24

Ten Tips, Tricks, and Shortcuts

A *tip* is a small suggestion, something spoken from experience or insight and something you may not have thought of yourself. A *trick* is something that's good to know — something surprising. And a *shortcut* is convenient and quick, like using duct tape as a parenting tool.

Although I'd like to think that everything I mention in this book is a tip, trick, or shortcut for using your Android phone, I can offer even more information. This chapter provides ten tips, tricks, and shortcuts to help you get the most from your phone.

Quickly Switch Apps

Android apps don't quit. Sure, some of them have a Quit or Sign Out command, but most apps lurk inside the phone's memory while you do other things. The Android operating system may eventually kill off a stale app. Before that happens, you can deftly and quickly switch between all running apps.

The key to making the switch is to use the Recent navigation icon, found at the bottom of the touchscreen. Figure 24-1 illustrates two popular representations of the Recent navigation icon.

FIGURE 24-1:
Incarnations
of the Recent
navigation icon.

Android Nougat
Android Marshmallow
Android Lollipop

Android Kitkat

After tapping the Recent navigation icon, choose an app from the list. Swipe the list up or down to peruse what's available. To dismiss the list, tap the Back or Home navigation icon.

>> On older phones that lack the Recent icon, long-press the Home navigation icon.

>> To remove an app from the list of recent apps, swipe it left or right. This is almost the same thing as quitting an app.

>> The list of recent apps is called the *Overview,* although everyone I know calls it the list of recent apps.

Instant Flashlight

One of the first "killer apps" on smartphones was the flashlight. It used the phone camera's LED flash to help you to see in the dark. For a while, everyone had to get a flashlight app, but today the flashlight feature is frequently found on the Quick Actions drawer.

Pull down the notifications drawer and pull again (if necessary) to display the quick actions. If one of them is a Flashlight icon, similar to the one shown in the margin, tap it to activate the flashlight feature. Tap again to turn off the flashlight.

>> The flashlight feature might also appear as a lock screen app, or you can place it on the lock screen as an option. Refer to Chapter 20.

>> Be aware that keeping the phone's LED lamp on for extended durations seriously drains the battery.

WARNING

>> Over time, it was discovered that the seemingly innocent flashlight apps actually spied on their phone owners. The apps would collect data and beam it back to a remote server somewhere. These apps were purged from the Google Play app library, but in any event the availability of the Quick Actions feature has rendered such apps unnecessary.

The Camera App's Panorama Mode

Most variations of the Camera app sport a panoramic shooting mode. The *panorama* is a wide shot — it works by panning the phone across a scene. The Camera app then stitches together several images to build a panoramic image.

To shoot a panoramic shot, follow these steps in the Google Camera app.

1. **Choose the Camera app's Panorama mode.**

   For the Google Camera app, tap the Side Menu icon to view shooting modes. Tap the Panorama icon, shown in the margin, to switch the app into Panorama mode.

   On a Samsung phone, tap the MODE button and choose Panorama.

2. **Hold the phone steady.**

3. **Tap the Shutter icon.**

4. **Pivot in one direction, following the onscreen animation.**

   Watch as the image is rendered and saved.

TIP

Panoramas work best for vistas, wide shoots, or perhaps for family gatherings where not everyone likes everyone else.

Avoid Data Surcharges

An important issue for everyone using an Android phone is whether you're about to burst through your monthly data quota. Mobile data surcharges can pinch the wallet, but your Android phone has a handy tool to help you avoid data overages. It's the data usage screen, shown in Figure 24-2.

FIGURE 24-2:
Data usage.

To access the data usage screen, follow these steps:

1. **Open the Settings app.**

2. **Choose Data Usage.**

   On some Samsung phones, choose the Connections tab at the top of the Settings app screen to find the Data Usage item.

The data usage screen sports two tabs: one for cellular (mobile data) access and another for Wi-Fi. If you don't see both tabs, tap the Action Overflow and choose Show Wi-Fi.

Each tab, Cellular and Wi-Fi, graphically illustrates data usage over time. As you can see from Figure 24-2, my phone uses far more Wi-Fi than mobile data, which is fine by my wallet.

The red and black lines on the Cellular tab can help remind you of how quickly you're filling the phone's mobile data quota. A warning message appears when data usage crosses the grey line. When the red line is reached, mobile data usage stops.

To set a warning (black line) limit, drag the tab on the right side of the screen up or down.

To set the stop (red line) limit, activate the Set Cellular Data Limit option; slide the Master Control icon to the On position. Then adjust the red line up or down.

To review access for a specific app, swipe the screen from bottom to top to view the App Usage area. Only apps that access the network appear. Tap an app to view detailed data usage information. If you notice that the app is using more data than it should, tap the App Settings button. You may be able to adjust the app's Internet access, which also helps avoid data surcharges.

Make the Phone Dream

Does your phone lock, or does it fall asleep? I prefer to think that the phone sleeps. That begs the question of whether or not it dreams.

Well, of course it does! You can even see the dreams, if you activate the Daydream feature — and if you keep the phone connected to a power source or in a docking station. Heed these steps:

1. **Start the Settings app.**

2. **Choose Display and then Daydream.**

   The Display item is found on the Device tab on certain Samsung phones.

3. **Ensure that the Daydream master control is in the On position.**

4. **Choose which type of daydream you want displayed.**

   The Clock is a popular item, though I'm fond of Colors.

 Some daydream choices feature the Settings icon, which is used to customize the daydream's appearance.

5. **Tap the Action Overflow and choose When to Daydream.**

6. **Choose the Either option.**

The daydreaming begins when the screen would normally time-out and lock, although the phone has to be receiving power. So, if you've set the phone to lock after 1 minute of inactivity, it daydreams instead — provided that it's plugged in or docked.

>> To disrupt the dream, swipe the screen.

>> The phone doesn't lock when it daydreams. To lock the phone, press the Power/Lock key.

>> Not every Android phone offers the Daydream feature.

Charge the Battery Without Wires

When you want to go truly wireless (I mean, forgo all wires — period), take advantage of wireless charging. This can be a phone feature, or you can add a wireless-charging battery cover to upgrade your phone.

After the phone is equipped for wireless charging, you simply set it on the charging pad or into the charging cradle. The battery charges. Voilà! No more wires.

Add Spice to Dictation

I feel that too few people use dictation, despite how handy it can be — especially for text messaging. Anyway, if you've used dictation, you might have noticed that it occasionally censors some of the words you utter. Perhaps you're the kind of person who doesn't put up with that kind of s***.

Relax. You can lift the vocal censorship ban by following these steps:

1. **Start the Settings app.**

2. **Choose Language & Input.**

   On some Samsung phones, you'll find the Language and Input item on either the Controls tab or the General tab in the Settings app.

3. **Choose the Google Voice Typing item.**

4. **Disable the Block Offensive Words option.**

And just what are offensive words? I would think that *censorship* is an offensive word. But no. Apparently, only a few choice words fall into this category. I won't print them here, because the phone's censor retains the initial letter and generally makes the foul language easy to guess. D***.

Add a Word to the Dictionary

Betcha didn't know that your phone sports a dictionary. The dictionary keeps track of words you type — words that it may not recognize as being spelled properly.

Words unknown to the phone are highlighted on the screen. Sometimes, the word is shown in a different color or on a different background or even underlined in red. To add that word to the phone's dictionary, long-press it. You see the Add Word to Dictionary command, which sticks the word in the phone's dictionary.

To review or edit the phone's dictionary, follow these steps:

1. **Start the Settings app.**

2. **Choose Language & Input.**

3. **Choose Personal Dictionary.**

   The command may not be obvious on some phones: Try choosing the keyboard first, and then choose either the Dictionary or User Dictionary command.

When the dictionary is visible, you can review words, edit them, remove them, or manually add new ones. To edit or delete a word, long-press it. To add a word manually, tap the Add icon.

Employ Some Useful Widgets

Your phone features a wide assortment of widgets with which to festoon the Home screen. They can be exceedingly handy, though you may not realize it because the sample widgets often included with the phone are tepid and lack inspiration.

Good widgets to add include navigation, contact, eBook, and web page favorites. Adding any of these widgets starts out the same. Here are the brief directions, with more specifics offered in Chapter 19:

1. **Long-press a Home screen page that has room for the widget.**

2. **Choose Widgets.**

3. **Drag the desired widget to the Home screen.**

4. **Complete the process.**

   The process is specific for each type of widget suggestion in this section.

Contacts/Direct Dial widget

Use the Direct Dial widget in the Contacts category to access those phone numbers you dial all the time. After adding the widget, choose a contact from the phone's address book. For contacts with multiple numbers, choose a number. (All are displayed in the list.) Tap this widget to dial the number instantly.

Maps/Directions widget

The Directions widget (in the Maps category) allows you to rapidly summon directions to a specific location from wherever you happen to be. After you plop the widget on the Home screen, select a traveling method and destination. You can type a contact name, an address, a business name, and so on. Add a shortcut name, which is a brief description to fit under the widget on the Home screen. Tap the Save button.

Tap the Directions widget to use it. Instantly, the Maps app starts and enters Navigation mode, steering you from wherever you are to the location referenced by the widget.

Google Play Books/Book widget

When you're mired in the middle of that latest potboiler, put a Google Play Books widget on the Home screen: Choose the Book widget (from the Google Play Books category), and then browse your digital library for the eBook you want to access.

Tap the widget to open the Play Books app and jump right to the spot where you were last reading.

Web page bookmark widget

While slapping down a web page bookmark widget would be an obvious way to keep handy a favorite web page shortcut on the Home screen, I have an easier way: Open the web browser app, navigate to the page you desire, and then tap the Action Overflow and choose Add to Home screen.

Take a Screen Shot

A *screen shot*, also called a *screen cap* (for *capture*), is a picture of your phone's touchscreen. If you see something interesting on the screen, or if you simply want to take a quick pic of your phone life, take a screen shot.

The stock Android method of shooting the screen is to press and hold both the Volume Down and Power/Lock keys at the same time. Upon success, the touchscreen image reduces in size, you may hear a shutter sound, and the screen shot is saved.

>> Open the Photos app to view your screen shots. Tap the Side Menu icon and choose Device Folders from the navigation drawer. Screen shots are saved in the Screenshots folder.

>> Some Samsung phones use a motion action to capture the screen: Hold your hand perpendicular to the phone, like you're giving it a karate chop. Swipe the edge of your palm over the screen, from right to left or from left to right. Upon success, you hear a shutter sound. The screen is captured.

TECHNICAL STUFF

>> Internally, screen shots are stored in the Pictures/Screenshots folder. They're created in the PNG graphics file format.

Chapter 25

Ten Things to Remember

I f only it were easy to narrow to ten items the list of all the things I want you to remember when using your Android phone. Even though you'll find in this chapter ten good things not to forget, don't think for a moment that there are only ten. In fact, as I remember more, I'll put them on my website, at wambooli. com. Check it for updates about your phone and perhaps for even more things to remember.

Lock the Phone on a Call

Whether you dialed out or someone dialed in, after you start talking, lock your phone. Press the Power/Lock key. By doing so, you disable the touchscreen and ensure that the call isn't unintentionally disconnected.

Of course, the call can still be disconnected by a dropped signal or by the other party getting all huffy and hanging up on you. But by locking the phone, you prevent a stray finger or your pocket from disconnecting (or muting) the phone.

TIP

If you like to talk with your hands, or use your hands for other tasks while you're on the phone (I sweep the floor, for example), get a good set of earbuds with a microphone. Using a headset lets you avoid trying to hold the phone between your ear and shoulder, which could unlock the phone or cause you to drop it or perhaps do something more perilous.

Switch to Landscape Orientation

The natural orientation of the Android phone is vertical — its portrait orientation. Even so, that doesn't mean you have to use an app in portrait orientation.

Rotating the phone to its side makes many apps, such as the web browser app and the Play Books app, appear wider. It's often a better way to see things in the phone's landscape orientation. For typing, that orientation gives you larger key caps on which to type when you use the onscreen keyboard.

REMEMBER

>> Record video in landscape orientation. If you can remember to do that, you'll find the thanks and appreciation of everyone who watches your movie.

>> Not every app supports landscape orientation.

>> The phone's Home screen may not support landscape orientation.

>> You can lock the orientation so that the touchscreen doesn't flip and flop. Refer to Chapter 3 for information on the screen rotation Quick Setting.

Dictate Text

Dictation is such a handy feature — don't forget to use it! You can dictate most text instead of typing it. Especially for text messages, it's just so quick and handy: Tap the Dictation (Microphone) key on the onscreen keyboard. Actually, anywhere you see the Dictation icon (shown in the margin), tap and begin speaking. Your utterances are translated to text. In most cases, the translation is instantaneous.

>> Refer to Chapter 4 for more information on dictation.

>> Google Now doesn't require you to tap the Microphone icon. Instead, utter the phrase "Okay, Google" and it starts listening. See Chapter 15 for information about Google Now.

Enjoy Predictive Text

Don't forget to take advantage of the suggestions that appear above the onscreen keyboard while you type. Tap a word suggestion to "type" that word. Plus, the predictive-text feature may instantly display the next logical word for you.

When predictive text fails you, keep in mind that you can use Google Gesture typing instead of the old hunt-and-peck. Dragging your finger over the keyboard and then choosing a word suggestion works quickly — when you remember to do it.

Refer to Chapter 20 for information on activating these keyboard settings.

Avoid Battery Hogs

Three items on your phone suck down battery power faster than an 18-year-old fleeing the tyranny of high school on graduation day:

>> The display

>> Navigation

>> Wireless radios

The display is obviously a most necessary part of your phone, but it's also a tremendous power hog. The Adaptive Brightness (also called Auto Brightness) setting is your best friend for saving power with the display. Refer to Chapter 20.

Navigation is certainly handy, but the battery drains rapidly because the phone's touchscreen is on the entire time, the voice is dictating directions, and the Wi-Fi and GPS radios may be active. If possible, plug the phone into the car's power socket while you're navigating. If you can't, keep a sharp eye on the battery meter.

Bluetooth requires extra power for its wireless radio. When you need that level of connectivity, great! Otherwise, turn off the phone's Bluetooth radio to save power.

Wireless radios include Wi-Fi networking, Bluetooth, and GPS. Though they do require extra power, they aren't power hogs, like navigation and the display. Still, when power is getting low, consider disabling these items.

Refer to Chapter 23 for more information on managing the phone's battery.

Beware of Roaming

Roaming can be expensive. The last non-smartphone (dumbphone?) I owned racked up $180 in roaming charges the month before I switched to a better cellular plan. Even though you might have a good phone plan, keep an eye on the phone's status bar to ensure that you don't see the Roaming status icon when you're making a call.

Well, yes, it's okay to make a call when your phone is roaming. My advice is to remember to check for the icon, not to avoid it. If possible, try to make your phone calls when you're back in your cellular service's coverage area. If you can't, make the phone call but keep in mind that you will be charged roaming fees. They ain't cheap.

Use the Plus (+) Symbol When Dialing Internationally

That phone number may look like it needs the + symbol, and the Phone app's dial-pad features a + key, shared with the 0 key — but don't use it unless you're dialing an international number. The + symbol prefix is the first part of any international phone number.

Refer to Chapter 22 for more information on international dialing.

Check Your Schedule

The Calendar app reminds you of upcoming dates and generally keeps you on schedule. A great way to augment the calendar is to employ the Calendar widget on the Home screen.

The Calendar widget lists the current date and then a long list of upcoming appointments. It's a great way to check your schedule, especially when you're a busy person and you occasionally forget to show up where you're needed.

Refer to Chapter 20 for information on adding widgets to the Home screen; Chapter 15 covers the Calendar app.

Snap a Pic of That Contact

Here's something I always forget: Whenever you're near one of your contacts, take the person's picture. Sure, some people are bashful, but most folks are flattered. The idea is to build up the phone's address book app so that all contacts have photos. Receiving a call is then much more interesting when you see the caller's picture, especially a silly or an embarrassing one.

REMEMBER

When taking the picture, be sure to show it to the person before you assign it to the contact. Let them decide whether it's good enough. Or, if you just want to be rude, assign a crummy-looking picture. Heck, you don't even have to do that: Just assign a random picture of anything. A plant. A rock. Your dog. But seriously, the next time you meet up with a contact, keep in mind that the phone can take that person's picture.

Refer to Chapter 13 for more information on using the phone's camera.

Use the Search Icon

Google is synonymous with searching, so you would expect the Google phone to sport robust searching abilities. And it does!

The Search icon is found in just about every Android app and on almost every screen on your phone. Tap that icon to look for information such as locations, people, text — you name it. It's handy. It's everywhere. Use it.

TECHNICAL STUFF

Once upon a time, the Search icon was one of the standard navigation icons that appeared at the bottom of every Android phone's touchscreen. The original four navigation icons were Home, Back, Menu (now Action Overflow), and Search.

Index

Symbols and Numerics

, (comma), 67
– (minus), 96
+ (plus), 285–286, 318
; (semicolon), 67
/ (slash), 49
1X, 226
3G, 226
4G LTE, 226

A

Aa icon, 206
abbreviations, in text messaging, 104
accessing
  address book, 86–88
  apps, 42
  data usage screen, 308
  Google Play, 212
  Maps app, 154
  Play Books app, 204
  Play Music notification, 188
  Quick Settings, 41–42
  slideshows, 175
  special characters, 50–52
  texting apps, 100
  Wi-Fi overseas, 290
accessories, adding, 19
accounts
  adding, 26–28
  email, 110–113
Action Bar icon, 45
Action Overflow icon, 45, 121, 127
activating
  Bluetooth on phone, 234–235
  dictation, 269
  for first time, 22–23
  gesture typing, 53, 269
  location technologies, 156
  predictive text, 268
  Wi-Fi, 227

adapters, for microSD card, 14
Add icon, 45
Add Photo icon, 141
Add Video icon, 141
adding
  accessories, 19
  accounts, 26–28
  contact pictures, 94–96
  contacts from call log, 90–91
  corporate email accounts, 113
  email accounts, 111–113
  email accounts manually, 112–113
  launchers to favorites tray, 251–252
  launchers to Home screen, 250–251
  layers in Maps app, 156
  music, 190–191
  owner info text, 280
  pauses when dialing numbers, 67
  people to hangouts, 145
  print service, 242–243
  songs to playlists, 194
  widgets to Home screen, 252–253
  words to dictionary, 311
address book. See also email; text messaging
  about, 85
  accessing, 86–88
  adding contact pictures, 94–96
  adding contacts from call log, 90–91
  creating contacts, 89–92
  creating contacts from email messages, 91–92
  creating favorites, 96
  editing, 93
  importing contacts from computer, 92–93
  joining contacts, 96–97
  managing, 93–98
  removing contacts, 98
  searching contacts, 89
  separating contacts, 97–98
  sorting, 88–89
addresses, finding with Maps app, 160–161
air travel, 287–288

items
  obtaining from Google Play, 214–216
  selecting groups of, 33
  sharing from Google Play, 220–221

J

joining contacts, 96–97
JPEG file format, 120, 169

K

keyboards
  customizing settings, 267–269
  generating feedback, 268
  onscreen, 47–49
  switching, 50

L

Labyrinth (game), 36
landscape orientation, switching to, 316
Last Added playlist, 192
launchers
  adding to favorites tray, 251–252
  adding to Home screen, 250–251
  defined, 250
  moving on Home screen, 254
  removing from Home screen, 255
layers, adding in Maps app, 156
LED flash, 16
LG, 18, 300
Linux operating system, 26
live wallpapers, 264
Location icon, 155
Location Tags feature, 175–176
location technologies, activating, 156
locations
  finding for pictures, 175–176
  finding with Maps app, 158–159
  sending from Maps app, 159–160
lock screen
  about, 24–25
  controlling notifications, 279
  customizing settings, 265–266

locking
  about, 29
  phones during calls, 316
Lollipop, 3
long-press, 32, 128
lost phones, 280–281

M

Mac, importing contacts from a, 92
maintenance and troubleshooting
  about, 291
  app issues, 298
  app support, 300
  backing up, 292–293
  battery, 294–297, 301
  cellular support, 299
  cleaning, 292
  connection issues, 297–298
  general issues, 297
  Google Play support, 300
  Help app, 299
  manufacturer support, 300
  music issues, 298
  orientation, 302
  overheating, 302
  regular maintenance, 292–293
  touchscreen, 301
  updating system, 293
Manage icon, 74
managing
  address book, 93–98
  apps, 256–260
  attachments in email, 119–120
  battery performance, 296–297
  files, 247
  Home screen pages, 264–265
  images, 178–182
  lock screen notifications, 279
  missed calls, 78–79
  multiple calls at once, 72–74
  multiple web pages, 131
  music, 187–188
  screen lock timeout, 267
  text messages, 107–108

N

navigating, to destinations in Maps app, 164–166

navigation icons, 33–34

Navigation mode, exiting, 166

Next key, 48

non-lock, 25

notifications
  about, 37–38
  controlling for lock screen, 279
  dismissing, 39
  Facebook, 138
  hiding, 39
  reviewing, 39–41
  ringtones for, 40

O

Off setting, for flash, 171

offline maps, saving, 156–158

On Device icon, 221

On setting, for flash, 171

1X, 226

OneDrive, 239

onscreen keyboard, 47–49

operations, basic, 31–36

organizing music, 192–195

orientation
  about, 15
  adding accessories, 19
  for Calculator app, 199
  for camera apps, 168
  changing, 36
  components, 16–18
  earphones, 18
  landscape, 316
  for onscreen keyboard, 49
  for Play Books app, 205–206
  for shooting video, 170
  troubleshooting, 302

Overflow icon, 46

overheating, troubleshooting, 302

overseas travel, 289–290

owner info text, adding, 280

P

pages
  about, 263
  Home screen, 264–265

pairing, with Bluetooth peripherals, 235–236

Pandora Radio app, 196

panning, in Maps app, 154

Panorama mode (Camera app), 307

paperclip icon, 119

Password lock, 25, 274

passwords, applying, 276

Pattern lock, 25, 274

Pause icon, 170

pauses, adding when dialing numbers, 67

PDF file format, 120

performing factory data reset, 282

peripherals, Bluetooth, 235–236

perspective, in Maps app, 154

phone. See telephone

Phone app, 38

phone case, 19

Phone icon, 25

phone status, 38

Photos app
  about, 173–174
  finding locations of pictures, 175–176
  managing images, 178–182
  posting videos to YouTube, 177–178
  sharing images, 177
  starting slideshows, 175
  viewing photos and videos, 174–175

pictures
  adding for contacts, 94–96
  backing up, 178–179
  cropping, 180–181
  deleting, 170–171, 182
  editing, 179
  finding locations of, 175–176
  grabbing from web pages, 134
  managing, 178–182
  rotating, 181–182
  sharing, 177
  shooting, 169

telephone *(continued)*
  finding lost, 280–281
  forwarding calls, 74–76
  locking during calls, 316
  managing multiple calls at once, 72–74
  placing calls, 62–65
  receiving calls, 68–70
  rejecting calls, 70–71
  silencing your, 35–36
  transferring information from old, 28
  using speed dial, 66
Tethering status icon, 233
text
  about, 47
  dictating, 53–54, 316–317
  editing, 55–57
  finding on web pages, 132
  Google Voice typing, 53–55
  onscreen keyboard, 47–49
  selecting, 56–57
  typing, 49–53
text chatting, in Google Hangouts app, 145–146
text messaging
  about, 99–100
  common abbreviations in, 104
  compared with emails, 101
  composing messages, 102–103
  conversations in, 103
  forwarding messages, 105
  group messages, 103
  managing, 107–108
  multimedia, 105–106
  opening texting apps, 100
  receiving messages, 103, 105
  rejecting calls with, 70–71
  removing, 107
  sending to contacts, 87
  setting ringtones for, 107–108
  texting contacts, 100–102
3G, 226
Thumbs Up playlist, 192
time zone
  changing, 288
  setting for events in Calendar app, 204
Tip icon, 4

T-Mobile, 299
touchscreen
  about, 16, 17
  for camera apps, 168
  configuring, 267
  manipulating, 32
  troubleshooting, 301
transferring
  files from media cards, 242
  files via USB cables, 239–242
  information from old phones, 28
Trash (Delete) icon, 45, 117, 182
traveling with your phone, 286–289
troubleshooting and maintenance
  about, 291
  app issues, 298
  app support, 300
  backing up, 292–293
  battery, 294–297, 301
  cellular support, 299
  cleaning, 292
  connection issues, 297–298
  general issues, 297
  Google Play support, 300
  Help app, 299
  manufacturer support, 300
  music issues, 298
  orientation, 302
  overheating, 302
  regular maintenance, 292–293
  touchscreen, 301
  updating system, 293
TuneIn Radio app, 196
turning off
  about, 29–30
  automatic backup, 178–179
  Bluetooth, 235, 296
  mobile hotspots, 232
  ringtone, 271
  vibration options, 296
  Wi-Fi, 227
turning on
  Bluetooth on phone, 234–235
  dictation, 269
  for first time, 22–23

About the Author

Dan Gookin has been writing about technology for over 25 years. He combines his love of writing with his gizmo fascination to create books that are informative, entertaining, and not boring. Having written over 150 titles with 12 million copies in print translated into over 30 languages, Dan can attest that his method of crafting computer tomes seems to work.

Perhaps his most famous title is the original *DOS For Dummies*, published in 1991. It became the world's fastest-selling computer book, at one time moving more copies per week than the *New York Times* number-one bestseller (though, as a reference, it could not be listed on the *Times'* Best Sellers list). That book spawned the entire line of *For Dummies* books, which remains a publishing phenomenon to this day.

Dan's most popular titles include *PCs For Dummies, Word For Dummies, Laptops For Dummies,* and *Android Tablets For Dummies*. He also maintains the vast and helpful website www.wambooli.com.

Dan holds a degree in Communications/Visual Arts from the University of California, San Diego. He lives in the Pacific Northwest, where he enjoys spending time with his sons playing video games indoors while they enjoy the gentle woods of Idaho.

Publisher's Acknowledgments

Acquisitions Editor: Katie Mohr

Senior Project Editor: Paul Levesque

Copy Editor: Rebecca Whitney

Editorial Assistant: Serena Novosel

Sr. Editorial Assistant: Cherie Case

Production Editor: Tamilmani Varadharaj

Cover Image: kirill_makarov/Shutterstock